Exploring Mathematical Thought

S J Taylor

Ginn London
Cheshire Melbourne

© S J Taylor 1970

077007   ISBN 0602 21385 1

Published in Great Britain by
Ginn & Company Ltd, 18 Bedford Row, London WC1
and in Australia by
F W Cheshire Publishing Pty Ltd, 346 St Kilda Road, Melbourne 3004

Made and printed in Great Britain by R & R Clark Ltd, Edinburgh

PREFACE

Many educated people instinctively shrink from the mention of mathematics. Often such people believe that mathematics consists of carrying out laborious calculations or manipulations with symbols in which there is little understanding, the only object of the exercise being to obtain a correct answer. The further misconception that mathematics was somehow established a long time ago, but has recently been revolutionized with the introduction of 'modern mathematics' to schools, is fostered by the experience of parents who suddenly discover they can no longer help their children with homework. In fact, mathematics is a living, growing subject. It is the servant of the scientific world, since the thought models used by scientists to describe the time–space universe all have a logical structure which is often formalized in mathematical terms. But mathematics is also a creative Art. Research in mathematics is usually motivated by the desire to discover new theories and the relationships between known theories. Its value is largely subjective, and a true mathematician appreciates a theorem or a proof as 'exciting' or 'beautiful' or 'surprising'.

Just as an appreciation of music or paintings requires education and experience, so we cannot hope to understand the nature of mathematics without some effort. A full appreciation of mathematics can only be attained by a few people after a lifetime of devotion to the subject, but no person can consider himself educated without some measure of understanding of the basic ideas of mathematics. Our object in the present book is to take a voyage of exploration into the world of mathematical thinking.

The choice of topics to be discussed in this book is somewhat arbitrary, but we avoid going deeply into areas involving a lot of technical competence. This means that the book should be accessible to those whose formal mathematical training ended at the age of 14 or 15. Actually we use very little of a technical nature in our discussion—so the reader is not required to remember what he may have been compelled to learn in his youth. We make no pretence of an overall coverage of mathematical ideas, but rather investigate selected topics in sufficient depth to indicate their importance.

The book is intended for serious study by intelligent readers—students in sixth forms, colleges of education, and universities—who are not specializing in mathematics. Important mathematical ideas are deep and sometimes difficult, but they can be grasped by anyone with a capacity for logical thought who is willing to invest some time in thinking about them. Our objective, therefore, is to present some of the ideas which have proved important in the development of mathematics in the last 100 years.

Those whose formal mathematical training has continued to a more advanced level may also find this book useful. For there is a danger in 'not being able to see the wood for the trees'—that is, many people trained in the use of mathematical techniques fail to appreciate the true nature of mathematics; they have not stopped to think about the basic ideas behind the techniques they have learnt to use. There is very little overlap between the contents of this book and the mathematics usually taught in the sixth form, so the book will also prove illuminating to those who have specialized in mathematics, but lack an overall feeling for the nature of the subject.

A book of this sort clearly contains little that is original and owes a great deal to the books that have gone before. There are a few historical comments, but no attempt is made to give credit to the original sources. Apart from one appendix containing a proof we have not been able to find in the literature, we have resisted the temptation to give detailed proofs which are complicated. Instead, you are referred to specific books where the details can be found.

I thank my wife, my publishers and my friends for their advice on detailed presentation and choice of topics and for telling me which parts of the book required elucidation to make them more intelligible. I am grateful for all advice received, but must take full responsibility for any errors and inaccuracies which remain.

S J Taylor

London
May 1970

HOW TO READ THIS BOOK

This book is intended to communicate basic ideas. Communication is only possible with your co-operation. It is essential to read carefully, trying to understand each paragraph. New and difficult ideas can only be absorbed slowly and after intellectual effort, so do not be disappointed if you have to read some parts two or even three times before the meaning is clear.

There are examples, chosen to illustrate the basic ideas, which have been worked out in detail in the text. In addition most sections contain some exercises to be solved; the harder exercises are marked with an asterisk. Experience shows that it is much easier to understand if you actually work something out with pencil and paper. In fact, the best way to read and understand mathematics is to try to reformulate the main ideas in your own words: no serious mathematician ever reads without paper to write on beside his book. If you are using this book in a group, you will find it very helpful to discuss the ideas. For, if you have to explain an idea to someone else, you will understand it more clearly yourself.

The chapters are largely independent, though the first two should be read before any others are attempted, and Chapter 5 should be read before Chapter 6. Mathematicians use a lot of notation, so there is an index of symbols provided, as well as an index of important mathematical terms. Always look up a symbol or word which is not familiar before going on, to make sure you understand its meaning. In most chapters there is a slight progression in difficulty from the beginning to the end, so do not be too discouraged if there are some sections you do not fully understand.

The ideas in this book are intended not only to interest you but also to excite your imagination. To understand this book will be hard work, but it should be satisfying and enjoyable. Pleasant reading!

CONTENTS

Preface	iii
How to read this book	v
Chapter 1. Language and Logic	1
1.1 Introduction	1
1.2 What is a set?	2
1.3 Subsets and complements	5
1.4 Unions and intersections	8
1.5 Logic in the language of sets	10
1.6 Algebra of sets	13
1.7 The structure of a proof	15
Chapter 2. Relations and Functions	19
2.1 Introduction	19
2.2 Cartesian product	22
2.3 What is a function?	26
2.4 Types of function	29
2.5 Operations on functions	33
Chapter 3. Properties of Natural Numbers	37
3.1 Introduction	37
3.2 Ordering and well ordering	42
3.3 Mathematical induction	44
3.4 Divisibility properties	48
3.5 Prime numbers	51
3.6 Goldbach's conjecture	54
Chapter 4. What is a Real Number?	56
4.1 Introduction	56
4.2 Rational numbers	59
4.3 Decimals	64
4.4 System of real numbers R	70
4.5 Some properties of R	71
Chapter 5. The Regularity of Randomness	77
5.1 What is probability?	77

5.2	Combining probabilities	80
5.3	Independent events	84
5.4	A class experiment	89
5.5	What determines p?	93
5.6	Law of large numbers	94
5.7	Applications	96
	Appendix	97

Chapter 6. The Gambler's Ruin. 100
- 6.1 Gambling in a fair game 100
- 6.2 Strategy in an unfair game 104
- 6.3 Premium Bonds 107
- 6.4 Football pools 110
- 6.5 Expectation 113
- 6.6 Wisdom in gambling 115

Chapter 7. Topology, or the Shape of a Set . . . 118
- 7.1 The nature of mathematics 118
- 7.2 Euler's formula for polyhedra 119
- 7.3 Topological properties 124
- 7.4 Colouring a map 129
- 7.5 Fixed-point theorem 132

Chapter 8. Paradoxes of the Infinite 136
- 8.1 A whole equal to part of itself 136
- 8.2 Cardinal number 139
- 8.3 Comparing cardinals 146

Chapter 9. Over the Top: Going to the Limit . . . 152
- 9.1 Some paradoxes 152
- 9.2 Beating the enemy 155
- 9.3 Accountant's nightmare 160
- 9.4 Summing an infinite series 164
- 9.5 How to measure the size of a set . . . 168

Bibliography 173
Index of Terms 175
Index of Symbols 181

1 | LANGUAGE AND LOGIC

1.1 Introduction

One of the fundamental assumptions behind all human thought is that man's reasoning process is valid and right in some absolute sense. This does not mean that we always reason correctly, but rather that there are valid processes of argument which always lead from a given hypothesis to the same conclusion. In practice the difficulty lies, to a large extent, in the problem of expressing, in a form which is both intelligible to others and unambiguous, these processes of logical thought.

One can attempt to express valid logical argument by a succession of English sentences (or by any other language). In this case it is essential to have a clear understanding of the precise meaning of each sentence. This is not always easy, for if we are expressing a complicated thought in terms of words, we require a complicated sentence. For example, if I say 'All Mr Jones' cows have either measles or mumps,' what exactly do I mean (see Fig. 1.1). In order to test whether you understand this completely, see if you can formulate a sentence which is equivalent to saying that the sentence in inverted

FIGURE 1.1

commas is false. It is even more difficult to think through a sequence of logical statements. For example, consider the following quotation from Lewis Carroll: 'If all boiled, red lobsters are dead, and all boiled, dead lobsters are red, does it follow that all red, dead lobsters are boiled?' The reader should try to decide this before going on—we shall analyse it later in the chapter.

The advertising industry often relies on consumers failing to think precisely. For example, it does not require a great deal of effort to imagine an advertisement for a new toothpaste, 'Whitey', which consisted of a picture of some famous film star showing two rows of perfect white teeth with the caption 'Miss Desirable Heploren uses Whitey: why not try it?' There is widespread use in advertisements of pictures indicating that attractiveness to the opposite sex is associated with the use of a certain product. If the potential consumer were to think clearly about the advertisements of this type he would realize that the hypothesis presented by the picture did not lead logically to the conclusion that he should buy the product. When we discuss valid arguments later in this chapter we shall be able to see how specious is the logic behind this type of advertisement.

To assist clear logical thought the mathematician has introduced a new language—the language of set theory. This lends itself to precision and, by introducing symbols which can be thought of as 'shorthand', it also results in arguments being communicated more concisely. Our real purpose in this chapter will be to introduce this language of sets. If you have already studied sets it is likely that the emphasis has been on numerical problems, but now we are interested more in analysing the process of logical reasoning.

Exercise 1a

Spend an hour watching commercial television and make a note of the content of the advertising material. Decide which of the advertisements lead you to think that you ought to buy the product advertised, and analyse them to see if your conclusion is a logical one.

1.2 What is a set?

The English language has many so-called collective nouns—herd (of cattle), bunch (of flowers), crowd (of people), flock (of sheep),

WHAT IS A SET?

among others. We use these words when we want to think of a collection (see Fig. 1.2) of separate objects rather than of the individuals making up the collection. Instead of several different words the mathematician picks on one word and uses it irrespective of the nature of the collection being considered. We call the collection a *set* and the individual objects composing it are called *members* of the set,

FIGURE 1.2

or *elements* in the set. We use italic capital letters $A, B, M, X, Y, \ldots$ to denote sets, and small letters to denote elements. We also write $\in$ for 'belongs to'. Thus:

$$a \in A$$

is a short way of writing the statement 'The object a is an element in the set A', and we read it 'a belongs to A'. Suppose A denotes the set of people attending the match between Arsenal and Manchester United on 20 January 1984, and a denotes Mr John Smith of 427 Singapore Road, Manchester; then $a \in A$ conveys a lot of information in just three symbols.

If the number of elements in a set X is finite we can count them and label the elements $x_1, x_2, \ldots, x_n$ (note that the $\ldots$ we have written between x_2 and x_n stands for whatever elements in the counting come between these two elements). We then write:

$$X = \{x_1, x_2, \ldots, x_n\}$$

to mean that the set X is precisely the collection of n objects labelled $x_1, x_2, \ldots, x_n$. For example, we could number the pieces of fruit in a bowl, or the bricks in a pile of child's bricks and then list the members of the set of fruit or the set of bricks in this way.

The notation of curly brackets can also be used when the set we are considering consists of precisely those objects which satisfy some prescribed condition. The condition is then called the *defining sentence* of the set. Thus:

$$A = \{a \mid a \text{ is in Form 2B of Kennilwatch School}\}$$

can be read as 'A is the set of those pupils a (the elements) who are in Form 2B of Kennilwatch School'. In such formulae the vertical bar | occurring in $\{x \mid \ldots\}$ can always be read to mean 'such that'. This notation is more useful when the sentence defining the set is mathematical in form. For example:

$$B = \{(x, y) \mid x = y^2\}$$

stands for the set of pairs of numbers x, y such that x equals y^2. In this case, the set B contains an infinite number of elements so that they could not have been written down in order. If we are thinking of whole numbers x, y then the set B contains the pairs (1, 1), (2, 4), ..., (100, 10 000), ...

Note that we must distinguish in our mind between the element x and the set $\{x\}$, which has only one element, namely x, in it. This distinction has sometimes been exploited by legislators. For suppose the constitution of a country disallows the passing of a law which singles out a particular township for special treatment. The crafty legislators then pass the law which is to apply to 'all townships with a population between 1400 and 1450 at the census of 1970'—and the defining sentence (in inverted commas) is such that precisely one town satisfies the condition.

In order to avoid frequent exceptions it is convenient to consider a special set, denoted by $\emptyset$, which has no elements in it. $\emptyset$ is called the *empty* set or the *null* set. It is amusing (but not very useful) to note that we could define $\emptyset$ using a defining sentence as follows:

$$\emptyset = \{x \mid x \neq x\}.$$

Since no element x satisfies the condition of being different from itself, the set of those elements x satisfying $x \neq x$ cannot contain any elements!

We say that two sets A and B are equal and write:

$$A = B$$

if A and B contain precisely the same elements. This is equivalent to saying that:

(i) every element in A is an element in B;
(ii) every element in B is an element in A.

Sets with quite different defining sentences can be equal. For example, suppose A is the set of girl pupils in your school whose age is less than 20, and B is the set of girl pupils in your school whose age is at least 3, then presumably you will be able to check that $A = B$.

When the teacher calls the register and finds that it corresponds exactly to the pupils in the room, he is proving that the set X of names in his register is equal to the set Y of names of all the pupils present. Note that the order is irrelevant to this problem—the register will be arranged alphabetically, while the pupils may sit in any order they please. We have, for example:

$$\{1, 2, 3, 4, 5\} = \{5, 2, 1, 4, 3\}.$$

Exercise 1b

1. Show that $\emptyset = \{x \mid x \text{ is a multiple of } 7 \text{ and } 8 < x < 11\}$.
2. Show that, if x is a number:

$$\{x \mid x > x\} = \{x \mid x.0 = 2\}.$$

3. List the elements of the following sets:
   (a) $\{P \mid P \text{ is the name of a British Prime Minister between 1950 and 1970}\}$
   (b) $\{x \mid x^2 = 16\}$
   (c) $\{x \mid 2x + 1 = 0\}$
   (d) $\{x \mid (x-2)(x+1)(x+1000) = 0\}$
   (e) $\{x \mid 1/x = x\}$
   (f) $\{x \mid x^2 = -9\}$

Hint: The usual process of solving an equation can be thought of as finding the elements of the set defined by the equation.

1.3 Subsets and complements

Usually in any particular discussion only a certain kind of object is relevant. The collection containing these relevant objects is called the 'discourse set', or the 'universe of discourse', or in more mathematical language the *whole space*, denoted by S. For example, in a discussion involving people as the objects it could be appropriate, according to the context, to take for S one of the following sets:
   (i) set of United Kingdom citizens as of 1 January 1970;
   (ii) set of human beings alive on the planet earth at this moment;

(*iii*) set of human beings who have inhabited or will have inhabited the earth at some time past, present, or future.

Again, if we are discussing properties of points on a single line it would be appropriate to take for our whole space R, the set of points on an infinite line. In this case the elements of the set R are the points on the line. If we are doing geography we might be interested in the set S of points on the surface of the earth—which could be represented by pairs (a, b), where a denotes the latitude and b the longitude of the point.

Suppose in a space S we are interested in selecting some of the elements of S and thinking of these together as a new set A. That is, for each element x in S, we can decide whether x is in A or not in A. This set A, which contains only elements from the space S, is called a *subset* of S. For example, if S is the set of people in the human race and A is the set of people living in Birmingham on 1 January 1969, then A is a subset of S. We write:

$$A \subset S$$

as a shorthand for 'A is a subset of S' or 'A is contained in S'. Another way of saying the same thing is 'S contains A', which we write:

$$S \supset A.$$

A mathematician called Venn invented a pictorial method of representing subsets of a space S. The points inside a fixed rectangle represent the whole space S, and subsets like A are represented by

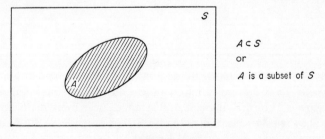

FIGURE 1.3

the points enclosed by some curve. Sometimes A is shaded in some distinct manner (see Fig. 1.3). We will find this method of representation, called a *Venn diagram*, useful in our further discussion of sets. More generally we say for a pair of sets A, B that:

$$A \subset B$$

if and only if every element in A is also an element in B. Since the empty set $\emptyset$ contains no element, it follows that any statement we make about the elements in $\emptyset$ is true. In particular, for any set A whatever, every element in $\emptyset$ is an element in A; so:

$$\emptyset \subset A$$

It is worth noting some other properties of the symbol $\subset$. Firstly note that, for any set A, $A \subset A$: this means that, for any set A, the sets $\emptyset$ and A are always subsets of A. Those subsets satisfying $B \subset A$ other than $B = \emptyset$ and $B = A$ are called *proper* subsets of A, to distinguish them from the trivial sets $\emptyset$ and A.

If A and B are two sets, the statement $A = B$ is equivalent to the two statements $A \subset B$ and $B \subset A$. You should go back to the definition of $\subset$ and $=$ for sets, and check that these deductions are valid.

We shall illustrate another property of the symbol $\subset$ with a Venn diagram. If A, B, C are any three sets such that $A \subset B$ and $B \subset C$, then $A \subset C$ (see Fig. 1.4).

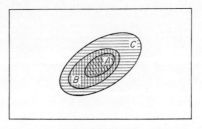

FIGURE 1.4

For a fixed space S, if A is a subset of S, then the set of those elements of S which are not in A is a written $(S - A)$; this set is called the *complement* of A in S. Note that the complement does depend on the space S. An alternative notation (which can be used when it is quite clear what space S we are considering) is $A' = S - A$. Thus if S is the set of pupils in Pugwash School and B denotes the boys of Pugwash, then $(S - B)$ or B' will be the set of girls in Pugwash.

Note that, in a given space S, for any two subsets A, B the statements $A \subset B$ and $B' \subset A'$ are equivalent; since every element in A

is an element in B if and only if every element of S which is not in B is not in A. For example, if C denotes the set of bullies in Pugwash School, then the statement $C \subset B$ (which is a short way of saying 'all the bullies are boys') is clearly the same as $B' \subset C'$ (which is a short way of saying that 'all the girls are not bullies').

Another important law of reasoning is that, in any fixed space S, for each subset A:

$$(A')' = A.$$

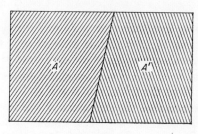

FIGURE 1.5

All we are really saying is that a subset A is only properly defined when it is known of each element in S whether it is in A or not in A. An element which is not in A' can only be in A (see Fig. 1.5).

Exercise 1c

1. In Pugwash School, find the complement of the set, B', of those pupils who are not boys.
2. Write down all the subsets of $S = \{1, 2, 3\}$ and find the complement of each of them.
3. How many subsets are there of each of the spaces $\{1, 2\}$, $\{1, 2, 3, 4\}$? Can you guess the number of distinct subsets of a space S which has n elements?
4. Draw a Venn diagram to illustrate $A \subset B$, and check that it also illustrates $B' \subset A'$.

1.4 Unions and intersections

If A, B are any two sets, the *union* of A and B denoted by:

$$A \cup B$$

(read 'A union B') is the set of elements which are in either A, or B, or both. The *intersection* of A and B denoted by:

$$A \cap B$$

(read 'A intersection B') is the set of elements which are in both A and B. These operations can be illustrated by Venn diagrams as in

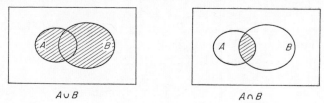

FIGURE 1.6

Fig. 1.6. It follows from the definitions, and is illustrated by the diagrams, that:

$$(A \cap B) \subset A \subset (A \cup B) \qquad (1.1)$$

and there is a similar relation with B in the middle. What we are saying in relation (1.1) is that any element which satisfies both the conditions for a set A and the conditions for a set B certainly satisfies the conditions for one of them. Further an element which is in one set, A say, satisfies the condition for the union $A \cup B$. If $A = \{1, 2, 4\}$, $B = \{2, 3, 4\}$, then $A \cup B = \{1, 2, 3, 4\}$, $A \cap B = \{2, 4\}$ and relation (1.1) is clearly satisfied.

Example 1. Suppose that in a given local election, individuals can only vote once in person at the polling station X and that there is an electoral roll containing the names of persons over 18 who reside in the constituency and persons over 18 who own a business in the constituency. If A is the set of names of residents over 18, and B is the set of names of businessmen over 18, then E, the electoral roll, is the set $A \cup B$ of names.

Note that there may be some names which are in both A and B, but they can only appear once in E. Further, if C is the set of persons who visit X to vote on polling day, then the set of people who actually vote is $C \cap E$, for you can vote only if you are there and your name is on the roll.

Exercise 1d

1. Draw a Venn diagram in which S is the human race, A is the set of residents of Birmingham and B is the set of all United Kingdom citizens. Express in terms of the sets A, B, A', B':
   (a) the set of UK citizens living in Birmingham;
   (b) the set of aliens—that is, those who are not UK citizens;
   (c) the set of aliens who live in Birmingham;
   (d) the set of residents of Birmingham together with all aliens.
2. Illustrate on a Venn diagram:
   (a) $A \subset B$ if and only if $A \cup B = B$;
   (b) $A \supset B$ if and only if $A \cap B = B$.
   Hint: in (a) draw a Venn diagram in which $A \subset B$ and show that $A \cup B = B$ follows. Then show in a diagram that if $A \cup B = B$, you must have $A \subset B$. These parts deal with 'if' and 'only if', respectively. Treat (b) in a similar manner.

1.5 Logic in the language of sets

We cannot speak about sets and operations on sets without using the structure of logical reasoning—and we have already used this freely without mention in this chapter. Let us now see how ordinary statements can be translated into set theory.

Example 2. 'George has red hair'. Here the obvious space to consider is the set S consisting of persons in the human race.

We can define a subset by:

$$R = \{p \in S \,|\, p \text{ has red hair}\}$$

(in words: R is the set of persons p in the human race such that p has red hair) and the statement becomes $g \in R$ (if g stands for George).

This is a typical example—for any property (here the property of having red hair) defines a set, namely the set of elements having this property.

Example 3. 'Everyone with red hair is bad-tempered.' We can take the same space S as in the last example, with the subset R as before, and a new subset:

$$B = \{p \in S \,|\, p \text{ is bad-tempered}\}.$$

The statement amounts to saying that every element in R is also in B or, in shorthand, $R \subset B$. Note that the two statements together imply that George is bad-tempered for $g \in R$ and every $p \in R$ satisfies $p \in B$, so $g \in B$. Draw a Venn diagram to illustrate this.

Example 1. 'All Mr Jones' cows have either measles or mumps.' Now the space S has to be the set of all cows:

$$M = \{c \in S \,|\, c \text{ has measles}\};$$
$$N = \{c \in S \,|\, c \text{ has mumps}\};$$
$$J = \{c \in S \,|\, c \text{ belongs to Mr Jones}\}.$$

Then $M \cup N$ is the set of cows with either measles or mumps (or both) and the statement can be written $J \subset M \cup N$.

We can now see easily what the negative of this statement is, for the statement is false if and only if we can find a cow c in J which is not in $M \cup N$—that is, which has not got measles and has not got mumps. In set language, $J \cap (M \cup N)' \neq \emptyset$. In words, the precise negative of the statement is 'Mr Jones has at least one cow which has neither measles nor mumps'.

Example 5. Think back to the advertisement for 'Whitey' on page 2. What is the real logical content of this?

The space S is again the human race and we need the following subsets:

$$B = \{p \in S \,|\, p \text{ is beautiful}\};$$
$$C = \{p \in S \,|\, p \text{ has good teeth}\};$$
$$W = \{p \in S \,|\, p \text{ uses 'Whitey'}\}.$$

The picture in the advertisement reminds us that, $x \in B \cap C$, where x is the film star, Desirable Heploren—for everyone believes that

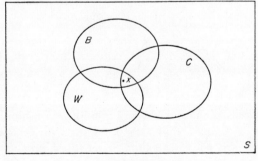

FIGURE I.7

$x \in B$ and $x \in C$. The caption of the advertisement gives us the further information that $x \in W$. If we illustrate this information by a Venn diagram (Fig. 1.7), all that we really know is that the sets B, C, W have at least one element, x, in common. The advertisement suggests the wrong deduction that in order to be in $B \cap C$ you have to be in W, or perhaps (even worse) that everyone in W is in $B \cap C$. Notice that, in logic, there is nothing to prevent a situation where x, the person in the advertisement, is the only element in $B \cap C$ even though the whole world is in W.

Example 6. Consider the Lewis Carroll riddle on page 2.

Now we must consider the whole space S to be the set of all lobsters, and again define three subsets:

$$R = \{x \in S \,|\, x \text{ is red}\};$$
$$D = \{x \in S \,|\, x \text{ is dead}\};$$
$$B = \{x \in S \,|\, x \text{ is boiled}\}.$$

The first two statements about lobsters are (*i*) $B \cap D \subset R$ and (*ii*) $B \cap R \subset D$, and we are asked whether (*iii*) $D \cap R \subset B$ follows.

Now clearly if there is at least one lobster which is red, dead and not boiled then (*iii*) will be violated, while (*i*) and (*ii*) will not be, since such an element will not be in the left hand side of either (*i*) or (*ii*). We can illustrate this possibility in a Venn diagram (see Fig. 1.8). In the diagram the areas ◭ and ▽ are part of the intersection of the two sets R and D, but these areas are not contained in the set B.

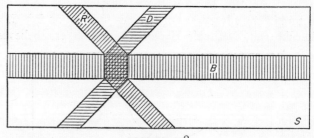

FIGURE 1.8

In fact the situation can be even worse—it is possible that none of the lobsters are boiled, while all of them are both dead and red for this situation does not violate either (*i*) or (*ii*). Thus (*iii*) is certainly not a valid deduction from (*i*) and (*ii*).

These examples illustrate the connection between logical statements and set theory. Before we can discuss further the methods of using this connection to make valid proofs, we need to obtain some further simple laws in set theory.

1.6 Algebra of sets

There are laws of reasoning that are relevant when we use compound sentences involving several clauses. These can be most clearly formulated using the laws of operations on sets. We consider these laws, and at the same time see their connection with our ordinary language structure.

Law 1 $\qquad (A')' = A$

Remember that in a given context there is a universe of discourse S, and A' denotes the complementary set to A—that is, the set of those elements which do not satisfy the condition defining A. The operation of taking the complement corresponds to the logical 'not'. Our law says that any statement which is not 'not true' is true—in logic two negatives make a positive.

The set operation of union $\cup$ corresponds to the logical 'or', and the set operation of intersection $\cap$ corresponds to the logical 'and'. In logic the word 'or' is not exclusive. Thus if we say that 'John has either measles or mumps', we mean that either John has measles or John has mumps or John has both measles and mumps. Sometimes use of the English word 'or' is intended to exclude the possibility 'both . . . and', whereas at other times this possibility is included. In logic we insist that 'or' is inclusive so that its meaning is precise.

Law 2 $\qquad A \cup B = B \cup A$
$\qquad\qquad\quad A \cap B = B \cap A$

These correspond to the fact that the order of clauses joined by 'or' or 'and' does not affect the meaning of a sentence. The meaning of the last sentence is unchanged if we write it as 'John has either mumps or measles'.

Law 3 $\qquad A \cup (B \cup C) = (A \cup B) \cup C$
$\qquad\qquad\quad A \cup (B \cup C) = (A \cup C) \cup B$

Here we are saying that the grouping of clauses, all connected by, say, 'or' makes no difference to the logical meaning. Thus 'John has measles or mumps or John has whooping cough' has the same meaning as 'John has measles, or John has mumps or whooping cough'—for both of these sentences mean that John is suffering from at least one of the three diseases mentioned.

Law 4 $\quad A \cup (B \cap C) = (A \cup B) \cap (A \cup C)$
$\quad\quad\quad\quad\quad A \cap (B \cup C) = (A \cap B) \cup (A \cap C)$

These are not quite so immediately obvious: they indicate clearly the precise meaning of sentences involving the two conjunctions 'and', 'or'. Thus the following pairs of sentences, illustrating the two laws, have the same meaning.
- (*i*) John has measles or he has both mumps and whooping cough. John has measles or mumps and in addition he has measles or whooping cough.
- (*ii*) John has measles and he has either mumps or whooping cough. John has measles and mumps or he has measles and whooping cough.

Law 5 $\quad (A \cup B)' = A' \cap B'$
$\quad\quad\quad\quad\quad (A \cap B)' = A' \cup B'$

These equations formalize the precise meaning of compound statements involving a logical 'not'. The opposite of 'John has either measles or mumps' is 'John has neither measles nor mumps', whereas the opposite of 'John has both measles and mumps' is 'either John is free from measles or he is free from mumps'.

The laws of set operation which we have formalized cannot really be proved. We should think of Laws 1 to 5 as an attempt to state concisely the basic rules of precise thought.

Exercise 1e

1. Draw a Venn diagram illustrating each of the Laws 2 to 5. Note that, though this shows that the Venn representation of the set operations agrees with the meaning of these laws, it cannot in any sense prove that the laws are valid.
2. In a regiment a soldier is asked to take over the duties of barber. His exact orders are to shave only those in the regiment who do not shave themselves. Should the soldier shave himself? If he

does then he is one who shaves himself and his orders direct him not to; however, if he does not then he is one who does not shave himself and his orders direct him to shave himself. Draw a Venn diagram illustrating this logical paradox. (For a discussion of the problem see page 42 of Rademacher and Toeplitz, 'The Enjoyment of Mathematics'.)

1.7 The structure of a proof

The object of a proof is to start with certain hypotheses and show that the processes of logical thought lead inevitably to a certain conclusion. In no sense can we be sure that the conclusion is true. All that we have proved is that 'if the hypothesis is true, then the conclusion is true'—and it is all too easy to use completely valid reasoning to prove nonsense, starting from a false hypothesis. In a proof one proceeds step by step, deducing (that is using the simple laws of reasoning we have formulated in terms of sets) statements one after another from the given hypothesis.

Let us introduce the shorthand '$\Rightarrow$' standing for 'implies', so that 'statement $\alpha \Rightarrow$ statement β' is another way of saying that 'if α is true, then β is true'. Laws 1 to 5 give us the basic rules of operating with 'and', 'or', 'not'. There are some more valid rules which we can write in shorthand and illustrate as follows:

(i) $A \subset B$, $B \subset C \Rightarrow A \subset C$; if all cats have whiskers, and any animal with whiskers can see in the dark, then all cats can see in the dark:

(ii) $x \in A$, $A \subset B \Rightarrow x \in B$; if Tiddles is a cat, and all cats have whiskers, then Tiddles has whiskers:

(iii) $A \subset B \Rightarrow B' \subset A'$; if all cats have whiskers, then any animal without whiskers is not a cat:

(iv) $A \cap B' = \emptyset \Rightarrow A \subset B$; if there is no cat which has not got whiskers, then all cats have whiskers.

A proof then consists of a chain of statements such that each statement of the chain is implied by those which precede it. The first statement of the chain (the hypothesis) then implies the last statement (the conclusion). It is the practice of mathematicians to formalize such proofs as *theorems*. Most theorems are of the type 'under certain specified conditions, the truth of statement α implies the truth of statement β'. In terms of set theory, the specified conditions determine the universe of discourse, or the whole space S; while the statements α, β determine subsets A, B of S. Then to prove the theorem we have to show that $α \Rightarrow β$, which is the same as showing that every element in the space S which satisfies the hypothesis α also satisfies the conclusion β—or that $A \subset B$.

There is more than one way of deducing logically that $A \subset B$. The most direct way is to form a 'chain' of statements starting from α and leading to β. But it is just as valid to start from 'not β' and form a chain leading to 'not α'. This means, under the specified conditions, the assumption that the conclusion β is false leads inevitably to the fact that the hypothesis α is false. A variant of this method is known as *reductio ad absurdum*. In this method we deduce a logical contradiction out of the assumption that α is true and β is false—in terms of set theory, a valid proof of $A \subset B$ is to show that $A \cap B' = \emptyset$.

To illustrate this *reductio ad absurdum* type of argument let us recall Euclid's beautiful proof that there is no largest prime number— or that the set of all prime numbers is infinite. Let us assume that we can do arithmetic on the whole numbers $1, 2, \ldots, n, \ldots$. As usual, we define a prime number to be a whole number which cannot be divided exactly (with no remainder) by any whole number other than itself and 1. This means that, if n is not a prime, then there are whole numbers a, b, neither of which is 1, such that $ab = n$. Now suppose there is a largest prime number p. Then we can write down all the prime numbers in order:

$$2, 3, 5, 7, \ldots, p,$$

and construct the whole number m equal to the product of all these primes plus 1, that is:

$$m = (2.3.5.7.\ldots.p) + 1.$$

Now m is not divisible by any of the primes $2, 3, \ldots, p$ since there is always a remainder 1 after such a division. Hence m is a prime, for if it had a factor other than itself or 1, this factor would have to be a prime or be divisible by some prime, and in either case m would be divisible by one of the primes. But obviously $m > p$, so p is not the largest prime. Thus our proof consists in showing that if p is the largest prime, then p is not the largest prime! (See pages 149–154 of Polya, 'How to Solve It', for a further discussion of this method of proof.)

One of the commonest mistakes in reasoning is to confuse $\alpha \Rightarrow \beta$ with $\beta \Rightarrow \alpha$. In mathematical terms this is the confusion between a theorem and its converse. Thus if $\alpha \Rightarrow \beta$ is a short way of saying that a certain hypothesis (summarized by α) implies the truth of a conclusion (summarized by β), the *converse* is the theorem which says that the hypothesis β implies the conclusion α. For example, we might be able to show that a family with an income in excess of £5000 per annum could afford to live in a house valued at £10 000, but this would not allow us to deduce that all houses valued at £10 000 were occupied by families with incomes in excess of £5000 per annum. In some cases both the theorem and its converse may be valid, but there is no reason to suppose that the converse is valid because the theorem is. Thus, 'every prime number bigger than 2 is odd' is valid; but 'every odd number bigger than 2 is prime' is false.

It is also important to distinguish clearly between a necessary and a sufficient condition. Thus, 'statement α is a *necessary* condition for the truth of β' means $\beta \Rightarrow \alpha$, but 'statement α is a *sufficient* condition for the truth of β' means $\alpha \Rightarrow \beta$. When we say that statement α is both necessary and sufficient for the truth of β, then we are claiming both implications $\alpha \Rightarrow \beta$ and $\beta \Rightarrow \alpha$. This corresponds to a theorem and its converse both being valid. For example $x = 1$ is a sufficient condition for the truth of the equation $x^2 = x$, but it is not a necessary condition. However, it is necessary, but not sufficient, for a whole number x to be even if it is to be divisible by 6. Finally, for a triangle to have two equal sides, it is necessary and sufficient for it to have two equal angles!

Once a theorem has been proved it becomes available for use in the proof of other theorems. This is the procedure whereby a body of mathematics is developed. There is a system of axioms—which are always assumed, and form part of the specified conditions determining the universe of discourse. A series of theorems is proved which explore the conclusions to be drawn from the axioms using other stated hypotheses.

Example 7. In manipulating equations it is very important to be clear that each line implies the next or that a given line is a necessary condition for the truth of the previous line. Lack of care easily produces nonsense. The following is a 'proof' that $2 = 1$:

Assume that	$a = b$
Multiply both sides by a:	$a^2 = ab$
Subtract b^2 from both sides:	$a^2 - b^2 = ab - b^2$
Factorize both sides:	$(a+b)(a-b) = b(a-b)$
Divide both sides by $(a-b)$:	$a + b = b$
Now put $a = b = 1$:	$2 = 1$

Exercise 1f

1. Where is the fallacy in the proof that $2 = 1$? (See chapter 5 of Northrop, 'Riddles in Mathematics', for a discussion of similar fallacies.)
2. Show that the following manipulation provides both a correct and an incorrect solution of the equation $\sqrt{x} + 1 = \sqrt{(4x)}$:

Square:	$x + 2\sqrt{x} + 1 = 4x$
Collect terms:	$2\sqrt{x} = 3x - 1$
Square:	$4x = 9x^2 - 6x + 1$
Collect terms:	$9x^2 - 10x + 1 = 0$
Factorize:	$(9x - 1)(x - 1) = 0$
Hence:	$x = \tfrac{1}{9}$ or $x = 1$

   Where is the mistake in the argument?

2 | RELATIONS AND FUNCTIONS

2.1 Introduction

If we say, in ordinary language, 'George is related to Mary', we have made a statement which is not quite precise: Fig. 2.1 shows two interpretations of the statement. Does it include the possibility that the relationship results from a marriage? If not, we might try to make

FIGURE 2.1

it precise by saying that George is related to Mary if, and only if, they have a common ancestor. Though this is a precise definition in the logical sense, it is too wide to cover normal usage, for we would not, in ordinary English, use the statement 'George is related to Mary' to cover the case of a common ancestor who lived, say, in Roman times.

In mathematical thought we insist on defining our terms and concepts precisely and unambiguously. The word 'relation' in mathematics will be reserved for some property which may or may not hold between two objects. The relation is only properly defined if it is such that, for each pair (x, y) of objects it is always possible to decide whether the stated relationship holds between x and y. Now, of course, by object we do not necessarily imply a physical object, but rather a mathematical object like a 'set' or an element from some specified set. In defining a relation precisely we shall have to say what kinds of objects x, y we wish to test to see whether x is related to y.

We can make our definition of relation clearer in the next section, but let us first consider some examples.

Example 1. We shall denote the relation 'is a brother of' by the letter ρ_1. If S denotes the set of all people, then we say that a male x is a brother of person y if and only if they have the same natural parents. This is uniquely defined, for we can decide of any pair x, y of persons in S whether or not x is the brother of y. We use the shorthand:

$$x \, \rho_1 \, y$$

for the statement 'x is a brother of y'. Note that, since only males can be brothers, we could also think of ρ_1 as a relation between elements x in M (the set of males) and elements y in S (see Fig. 2.2).

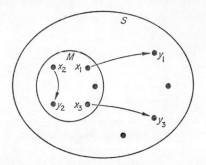

FIGURE 2.2 The relation ρ_1 (is the brother of) is represented by arrows

Example 2. Consider the relation 'is descended from': ρ_2. This is again well defined for pairs x, y from the set S of all people. Thus we write:

$$x \, \rho_2 \, y$$

if it is possible to find a chain of people $y = y_1, y_2, \ldots, y_n = x$ such that each person in the chain, y_i, is the natural mother or father of the next person, y_{i+1} (see Fig. 2.3).

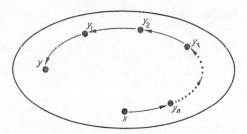

FIGURE 2.3 An arrow or a succession of arrows represents the relation ρ_2 (is descended from)

Example 3. Consider the relation 'is the square of': ρ_3. This relation has nothing to do with the set of people: it can be defined on any set for which multiplication can be carried out. For example, if N is the set of whole numbers, we say that for $x, y \in N$:

$$x \; \rho_3 \; y$$

if and only if $x = y^2$. In Fig. 2.4, ρ_3 is represented by arrows. (Note that $x, y \in N$ is the notation for 'x and y are elements of (or are in) the set N'.)

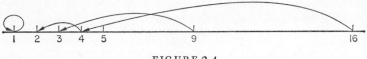

FIGURE 2.4

Example 4. Consider the relation 'is greater than': ρ_4. This can again be thought of as a relation on N. We say that a pair x, y in N satisfies:

$$x \; \rho_4 \; y$$

if the number x comes after the number y in the ordinary process of counting. (The process of counting will be examined in more detail in Chapter 3.)

Exercise 2a

Define precisely:
(a) 'is a cousin of'
(b) 'is a great-aunt of'
so that they become relations for pairs x, y of persons in the set S of all people.

2.2 Cartesian product

In our discussion of relation in the last section we used pairs x, y of elements sometimes from a single set, sometimes from two different sets. In each of the definitions for $x \rho y$ the order of the pair was important—for example, the statement that person x is descended from person y is very different from the statement that person y is descended from person x. This means that the objects we are considering are really *ordered pairs* of elements the first being from a set A, the second from a set B. We denote by:

$$A \times B$$

(read 'A cross B') the set consisting of all ordered pairs (a, b) where $a \in A$ and $b \in B$.

This set $A \times B$ is called the *product set* of A and B, or the *Cartesian product* of A and B. We call a the *first co-ordinate* of the ordered pair (a, b) and b the *second co-ordinate*.

Example 5. If $A = \{a, e, i, o, u\}$ and $B = \{x, y, z\}$ then the product set $A \times B$ has 15 elements in it, namely:

$(a, x), (a, y), (a, z), (e, x), (e, y), (e, z), (i, x), (i, y),$
$(i, z), (o, x), (o, y), (o, z), (u, x), (u, y), (u, z).$

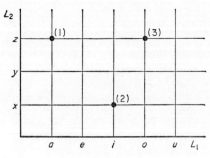

FIGURE 2.5

CARTESIAN PRODUCT

It is convenient to illustrate a product set by means of a co-ordinate diagram of $A \times B$. For this purpose we display the points of A on a horizontal line L_1, and those of B on a vertical line L_2. Lines are then drawn through each of the points of A in the vertical direction, and through each point of B in the horizontal direction. The intersections of these lines then denote points of $A \times B$. Thus in Fig. 2.5 (1) is the point (a, z), (2) is the point (i, x) and (3) is the point (o, z).

This process is often used in fixing the position of seats in a theatre. Suppose the school assembly hall is used for a play and the seats are arranged as shown in Fig. 2.6. Each seat is shown as a

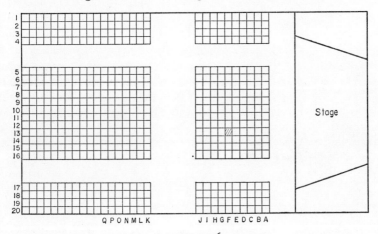

FIGURE 2.6

square, the row is given a letter, and on each row the seats are numbered from left to right. We can then easily find the seat (F, 13) which we have booked, for the system of designating the seats is precisely that of a Cartesian product.

Example 6. Suppose R denotes the set of real numbers (these will be discussed in Chapter 4, but we can think of the real numbers as representing points on a directed line). Then we denote by $R \times R$ or R^2 the set of ordered pairs of real numbers. Its co-ordinate diagram represents the geometry of the plane. Thus if x is the real number represented by a point on the line L_1, and y is the real number represented by a point on the perpendicular line, we obtain a representation of the point (x, y) in R^2 by finding the intersection of the line through x which is perpendicular to L_1, and the line through y which is parallel to L_1 (see Fig. 2.7).

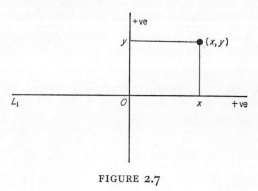

FIGURE 2.7

We can now say more precisely what we mean by a relation between elements of a set A and those of a set B. The examples which we considered in Section 2.1 all fit in with this precise definition.

Definition

Any subset of $A \times B$ is called a (binary) *relation* from A to B (or a relation in $A \times B$).

In this definition, A, B are any prescribed sets and the possibility that $A = B$ is in no way excluded. A relation in $A \times A$ is sometimes called a relation on A.

If E is a subset of $A \times B$, then we can tell of any ordered pair (x, y), where $x \in A$ and $y \in B$, whether or not (x, y) is an element of E. Then if $(x, y) \in E$, we say that x is in relation ρ_E to y, whereas if $(x, y) \notin E$ we say that x is not in relation ρ_E to y. Conversely any relation ρ must define a subset of $A \times B$, for we can say that $(x, y) \in E$ if and only if x is in relation ρ to y.

Example 7. In a certain tennis club the members are arranged in a 'ladder' or ranking which is intended to indicate that member x is better than member y in the ranking. If we illustrate this in a co-ordinate diagram, in which on each co-ordinate line persons are arranged in decreasing order of merit, then the marked points (see Fig. 2.8) are precisely those ordered pairs (x, y) such that x comes before y in the ranking. This illustrates the relation 'x is better than y at tennis'.

CARTESIAN PRODUCT 25

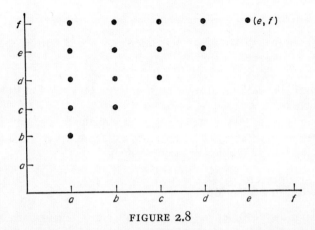

FIGURE 2.8

Example 8. The relation of Example 3 can be considered on $R \times R$. This will correspond to the subset of the plane consisting of points (x, y) such that $x = y^2$. If we represent $R \times R$ as in Fig. 2.7 we obtain Fig. 2.9 as a representation of this relation. The curve in this figure contains only points (x, y) such that x is the square of y.

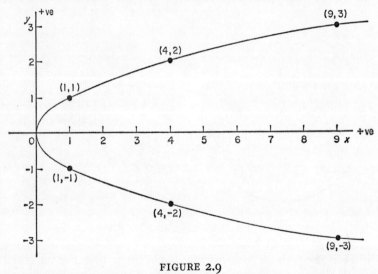

FIGURE 2.9

Note that the representation of a relation by means of a coordinate diagram is called the *graph* of the relation.

B

Exercise 2b

1. Show that the relations of Examples 1 and 2 can be obtained from subsets of a suitable Cartesian product.
2. Draw the graph in R^2 ($= R \times R$) of the relation defined in Example 4. This shows that the graph of a relation does not have to be a curve.
*3. For sets E, F the statement E is contained in F, or $E \subset F$ defined in Section 1.3, determines a relation. Interpret this as a subset of a suitable Cartesian product.

2.3 What is a function?

The notion of a function (otherwise called mapping or transformation) is absolutely central to all branches of modern mathematics. The basic idea of a function is that of one 'object' completely determining another. Thus if we construct a 'map' of a portion of the earth's surface, we are representing the portion on a piece of paper in such a way that each point x on the ground corresponds to a unique point $f(x)$ on the paper (see Fig. 2.10). In geography one learns about

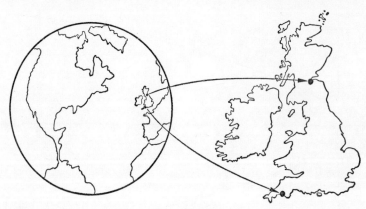

FIGURE 2.10

different methods of constructing such maps (the method used depends on the purpose for which the map is wanted), but in each such map $f(x)$ is uniquely determined by x.

Another example of this kind is the rule which assigns to each per-

WHAT IS A FUNCTION?

son p his or her natural mother $m(p)$. The rule assigning to each person his or her grandmother does not lead to a uniquely defined person and so does not satisfy our concept (see Fig. 2.11).

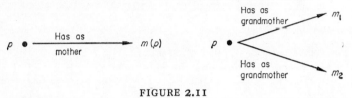

FIGURE 2.11

Definition

We say that we have a function f, defined on a set A, which gives elements in a set B if there is some rule f such that, for each element x in A, a single, uniquely defined element $f(x)$ in B is determined.

If we illustrate by an arrow our rule or correspondence which gets us from A to B, then typical correspondences between sets with four elements would be as shown in Fig. 2.12.

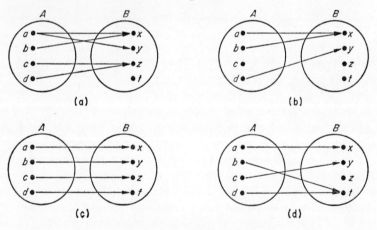

FIGURE 2.12

Figure 2.12a does not represent a function because a corresponds to two elements x and y, whereas Fig. 2.12b fails to represent a function because c does not correspond to an element of B. Figures 2.12c and **d** do represent functions, for there is one and only one arrow emanating from each point of A. Note that we do not require each element in B to correspond to at least one element from A, nor do we object if there are elements in B which come from two or more elements of A.

Exercise 2c

For sets $A = \{-2, -1, 0, 1, 2, 3\}$ and $B = \{0, 1, 2, \ldots, 9, 10\}$ consider the function defined by the rule 'take an element of A and square it'. Draw diagrams like Fig. 2.12 to illustrate this function and notice that some elements of B are taken twice by the function, while others cannot be taken at all.

We use the shorthand $f : A \to B$ for the statement 'f is a function defined on A taking values on B'. It is clear that a function f is a special kind of relation on $A \times B$. Figure 2.12a illustrates the relation given by the set:

$$(a, x), (a, y), (b, x), (c, z), (d, z).$$

It fails to be a function because the first co-ordinate a occurs twice in distinct ordered pairs. Similarly Fig. 2.12b fails to illustrate a function because the first co-ordinate c does not occur among the ordered pairs $(a, x), (b, x), (d, y)$. Thus a relation ρ from A to B is a function defined on A taking values in B if and only if:

(*i*) each element $x \in A$ is related to at least one element of B;

(*ii*) no element x in A is related to more than one element of B.

For, given $x \in A$, the rule $f(x)$ which defines $y \in B$ is to find an element y in B such that $x \rho y$. Condition (*i*) ensures that there is such a y and (*ii*) ensures that y is unique. Note that our definition of function is not symmetric; that is, we cannot interchange the roles of the two sets A, B. Figure 2.12d illustrates a relation which is a function from A to B, but not a function from B to A, since the element z in B does not correspond to any element in A and t corresponds to two elements.

It is instructive to examine the graph of a relation which is a function. Figure 2.13 illustrates the graph of a function from a 6-point set A to a 3-point set B. Then condition (*i*) ensures that each vertical line through a point of A must meet the graph, while condition (*ii*) ensures that each such line meets the graph precisely once. These two facts characterize the graph of a function.

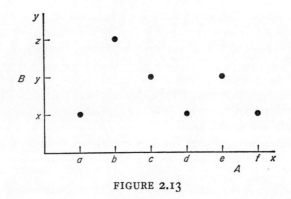

FIGURE 2.13

Example 9. The relation 'x is a brother of y' does not define a function on S to S, since (i) not all people have brothers, and (ii) some people have more than one brother.

Exercise 2d

1. Show that the rule which assigns to each person his or her natural mother is a function defined on S the set of all people taking values in F, the set of all females.
2. Show that the 'ladder' discussed in Example 7 does not define a function.
3. Show that the relation, whose graph is illustrated in Fig. 2.9, given by set of (x, y) such that $x = y^2$, $x \in R$, $y \in R$, does not determine a function from y to x.

2.4 Types of function

We have defined a function $f : A \to B$ to be a rule which assigns to each element $x \in A$ a unique element $f(x) \in B$. We consider two properties which may or may not be satisfied by a function:

(i) Are all points of B attained as an image $f(x)$ of the function f at some point of B? If for all $y \in B$, we can find an element $x \in A$ such that $y = f(x)$, we say that f maps A onto B.

(ii) Are all points of B attained at most once as an image $f(x)$ of a point in A? We say that the function $f : A \to B$ is (1, 1) (or *one-to-one*) if distinct points x_1, $x_2 \in A$ ($x_1 \neq x_2$) map into distinct points $f(x_1) \neq f(x_2)$ in B.

It is clear that Fig. 2.12c illustrates a function which is both (1, 1) and onto, Fig. 2.12d is a function which is neither (1, 1) nor onto, and Fig. 2.13 is the graph of a function which is onto but not (1, 1) (since $x \in B$ is the image of a, d, f in A). The condition that $f: A \to B$ maps A onto B is a restriction only on the size of B—it really says that B is just big enough to contain all the elements $f(x)$ for $x \in A$. However, the condition that $f: A \to B$ is (1, 1) only restricts the values taken by f at different points in A.

Sometimes we can alter the sets concerned to make a function (1, 1) or onto or both.

Example 10. $f: R \to R$ is defined by $y = f(x) = x^2$ for all real numbers x.

This is not 'onto' R since $f(x)$ is always positive or zero for any real number x—this means that there are no real numbers x such that $x^2 = -2$, for instance. However, if we denote by R^+ the set of real numbers x such that $x \geqslant 0$, then $f: R \to R^+$ defined by $y = f(x) = x^2$ is a function from R onto R^+ (see Fig. 2.14b). [It is worth observing that this is not a trivial statement, but rather a deep property of the real numbers to which we will return in Chapter 4.] However, our function $f: R \to R^+$ is clearly not (1, 1), for $f(-x) = (-x)^2 = x^2 = f(x)$: that is, the function takes the same value for two distinct elements; for instance, $f(-10) = f(10) = 100$.

If we now restrict the function so that it is defined on R^+ taking values in R^+ it becomes both (1, 1) and onto (see Fig. 2.14c), for now if $x_1 \neq x_2$ then one of them is larger, say $x_1 > x_2 > 0$. This implies $f(x_1) > f(x_2)$. Thus $f: R^+ \to R^+$ defined by $y = f(x) = x^2$ is both (1, 1) and onto. The graphs showing the relation $y = x^2$, but defined on alternative sets, are given in Fig. 2.14.

This example illustrates that the formula $y = x^2$ is not itself the

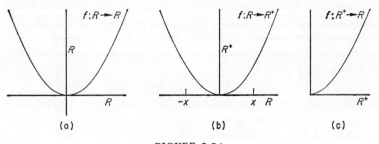

FIGURE 2.14

function. It is not correct therefore to speak of the function x^2, or indeed, of the function $f(x)$ where $f(x)$ is some explicit formula. A formula such as x^2 only defines a function when we have specified the set on which it is to be applied—called the *domain* of the function— and the set on which it takes its values—called the *range* of the function. We can then say $f: A \to B$ is a function such that each point x in A is mapped to the point $y = f(x)$ in B. It may be possible to express this rule f which is applied to point x in terms of a formula (such as x^2), or it may not.

Functions which are (1, 1) and onto are so important that we give them the special name *bijection*—or we say that the function is *bijective*.

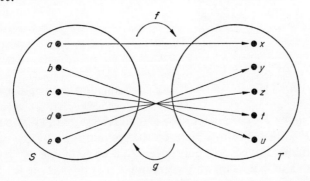

FIGURE 2.15

Figure 2.15 illustrates a function f defined on a set S with five members to a set T with five members which is (1, 1) and onto. If we reversed all the arrows in the diagram it is clear that we would have defined a function $g : T \to S$. This is typical of this situation. For suppose A, B are any sets (they may contain a finite or an infinite number of points) and $f : A \to B$ is a bijection. For each y in B we know there is an $x \in A$ such that $y = f(x)$ because f maps A onto B, and we know there is only one such x because f is a (1, 1) function. So if we denote by $g(y) = x$ the unique point x of A such that $f(x) = y$ we have defined a function g on B taking values in A. In this situation we call $g : B \to A$ the *inverse* function of $f : A \to B$.

Example 11. If E is that portion of the earth's surface called England, and we draw a map of England on a piece of paper, we mark a set F

on our paper with the property that to each point p on the ground in England, there corresponds a unique point $m(p)$ on our paper (see

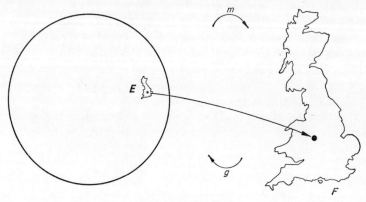

FIGURE 2.16

Fig. 2.16). Thus $m : E \to F$ is a function. But it is only a useful map if:
   (i) the set F drawn on our paper is all relevant, *i.e.*, every point in F corresponds to a point on the ground;
   (ii) the set F is covered only once, *i.e.*, each point in F corresponds to a unique point on the ground.
What we are really saying is that when we draw a map we insist that the mapping function $m : E \to F$ is a bijection. This means that there is an inverse function $g : F \to E$ so that each point on the paper in the set F corresponds to a unique point on the ground somewhere in England.

Example 12. Let N be the set of positive integers or whole numbers $\{1, 2, 3, \ldots\}$ and E be any set. Then a sequence $x_1, x_2, \ldots, x_n, \ldots$ of elements of E specifies particular elements of E in the order of the integers. What this really means is that we have a function $s : N \to E$ which assigns to each positive integer n in N a unique element $s(n)$ in E which we choose to label x_n. The precise mathematical definition of a *sequence* in E is just a function $S : N \to E$.

Exercise 2e

1. Suppose S is the set of all people and $m : S \to S$ is the function which assigns to each person p his natural mother $m(p)$. Show that this function is neither (1, 1) nor onto. Can you alter the domain to make it (1, 1), and the range to make it onto?
2. M is a set of 5 males and F is a set of 7 females. An arrow represents the relation 'is the husband of' (see Fig. 2.17). Does this determine a function? Could it be a function if:
   (a) A man commits bigamy?
   (b) A woman commits bigamy?

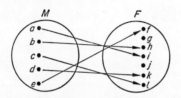

FIGURE 2.17

3. Premium Bonds are listed by means of a sequence of letters and digits. Show that the act of buying a premium bond defines a function from a suitable set to the set S of all people. This function is (1, 1) but not onto.
4. For a given function $f : A \to B$, write down a simple condition for:
   (a) the function f not to be (1, 1);
   (b) the function f not to map A onto B.
5. Consider the graph of any function $f : A \to B$. Show that the function will be (1, 1) if and only if each horizontal line through points of B meets the graph at most once, and the function is onto if and only if each horizontal line through points of B meets the graph at least once. This means that if $f : A \to B$ is a bijection, then the graph of the function f becomes the graph of the inverse function $g : B \to A$ if you turn the paper through a right angle.

2.5 Operations on functions

Let us illustrate the situation where we have two functions

$f: U \to V$, $g: V \to W$ such that the domain of the second function is the range of the first (see Fig. 2.18). If to each point in U we first

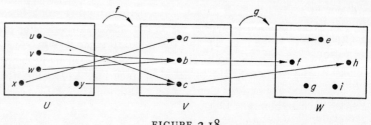

FIGURE 2.18

carry out rule f, giving us a point of V to which we apply rule g, we end up with a (unique) point in W. This means that the two functions $f: U \to V$, $g: V \to W$ determine completely a function $h: U \to W$ called the *composition* of f and g. If Fig. 2.18 illustrates the functions f, g, then Fig. 2.19 will illustrate the composition h which is sometimes denoted '$g \circ f$' (to be read 'g circle f' or 'g operating on f').

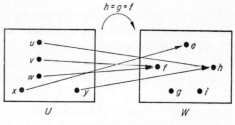

FIGURE 2.19

Example 13. Suppose I want to know someone's height in centimetres, but I have only got a tape measure calibrated in inches. I can measure the person in inches and then 'convert' the result to centimetres. The first operation of using the tape measure to determine the height of the person in inches is a function $f: S \to R$ where S is the set of all people, R is the set of real numbers. The conversion from inches to centimetres is a function $g: R \to R$ where, for each number x in R, we obtain y in R by using the formula $y = g(x) = c \cdot x$ where c is the number of centimetres in one inch ($c = 2 \cdot 54$). The result is a function $g \circ f: S \to R$ which assigns to each person his height in centimetres.

OPERATIONS ON FUNCTIONS

Example 14. The calculation of the amount of income tax a person has to pay is another example of this type of composition. The first stage is to determine a person's taxable income in £. (There is a very complicated set of rules to be applied to the individual which results in a final figure which takes account of all his circumstances as well as his various sources of income.) This is a function $f: S \to R$, even though it would not be possible to write down a single formula which could be used to calculate a person's taxable income. The second stage is to use the 'tax tables' to find the tax payable on the taxable income determined. This is a function $g: R \to R$ such that, if x is the taxable income in £, $y = g(x)$ is the amount of tax payable. Then

FIGURE 2.20

$g \circ f: S \to R$ is the function which determines how much income tax the person is liable to pay. Because of the system of tax collection, there is a final operation, namely to subtract the total amount of tax already collected from the person p from the figure $g \circ f(p)$ of the liability. If this results in a positive answer, the person p is then sent a bill for tax underpaid, if it results in a negative answer, then p can demand back the tax he has overpaid.

Example 15. Suppose $f: A \to B$ is a bijection. We have seen that the inverse function $g: B \to A$ can be defined. Then the composition $g \circ f: A \to A$ maps each point x in A into itself, for $g(y)$ is precisely the unique x such that $y = f(x)$. Similarly $f \circ g: B \to B$ maps each point y in B into itself.

Note that for any set E, the function $Id_E : E \to E$ which maps each point of E into itself is called the *identity function* on E. If you consider the function $f : R \to R$ defined by $f(x) = y = mx + c$, where m and c are fixed real numbers, the graph will be a straight line. The special case $m = 1$, $c = 0$ corresponds to the identity function Id_R. Its graph is the set of points (x, x) in $R \times R$. This means that if $g : B \to A$ is the inverse function of $f : A \to B$, then:

$$g \circ f = Id_A \qquad f \circ g = Id_B$$

One should think of the identity function on E as a 'boomerang' because each point $x \in E$ goes exactly to itself (see Fig. 2.20).

Exercise 2f

*Suppose $f : A \to B$, $g : B \to C$ are functions such that the composition $g \circ f : A \to C$ is (1, 1) and onto. Show that f is a (1, 1) function and g maps B onto C.

3 | PROPERTIES OF NATURAL NUMBERS

3.1 Introduction

The reader who disliked number work in the primary school, and subsequently became confirmed in his belief that arithmetic is a bore, should not give up in despair at the idea of a chapter on numbers. Our objectives here are philosophical rather than computational: we shall try to understand why the processes of arithmetic actually work, and we shall not require you to do any 'sums'.

For most people, learning to count is their first encounter with

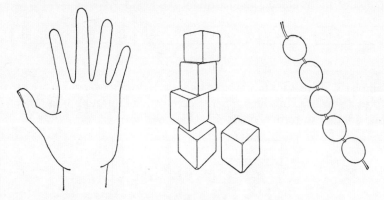

FIGURE 3.1

mathematics. Initially, a child thinks of a number, say five, as a property of a particular collection of beads or fingers. Then he realizes that a collection of five fingers, a collection of five beads and a collection of five blocks have something in common (Fig. 3.1). When each of these collections is counted, the same number 'five' results. The child learns to write a symbol '5' for the result of counting any collection of five objects, and eventually the number '5' begins to have some abstract significance in his mind.

Mathematically, this connection between the number '5' and different sets of five elements can be formalized using the idea of a one-to-one correspondence, or a function from a set A to a set B which is (1, 1) and onto. We will examine this idea in Chapter 8, but for the moment we assume that your intuition about the 'number' of oranges in a bowl or the number of elements in a finite set is valid.

The numbers 1, 2, 3 . . . which we get when we count are called *natural numbers* or *integers*, and we denote the set of all such numbers by N. An element $a \in N$ can be illustrated by a rectangular box with the appropriate number of dots in it, as in Fig. 3.2. The child soon

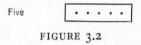

FIGURE 3.2

learns to add two numbers by putting two collections together and counting them as one. We can illustrate this by putting the boxes together and removing the partition (see Fig. 3.3):

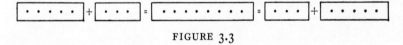

FIGURE 3.3

Similarly, if we want to multiply two whole numbers, a and b, we can form a new box, in which a boxes each with b dots are put on top of one another to form a single pile (Fig. 3.4):

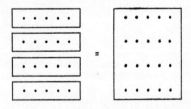

FIGURE 3.4

The child eventually realizes that the arithmetic operations of addition and multiplication always conform to certain rules. Hence, if $a, b, c, \in N$, we write $a + b$ for the sum of a and b, and $a.b$ or ab for the product of a and b. The rules can be formulated as follows:

INTRODUCTION 39

Law 1	$a+b = b+a$	Commutative law of addition
Law 2	$ab = ba$	Commutative law of multiplication
Law 3	$(a+b)+c = a+(b+c)$	Associative law of addition
Law 4	$(ab)c = a(bc)$	Associative law of multiplication
Law 5	$a(b+c) = ab+ac$	Distributive law
Law 6	$1.a = a$	Integer 1 is a unit for multiplication

Note that Fig. 3.3 illustrates Law 1 for $a = 5$, $b = 3$. Turning the large rectangle in Fig. 3.4 through a right angle, results in an illustration of Law 2 for $a = 4$, $b = 5$. These commutative laws seem obvious when we add or multiply integers, but it is worth noticing that there are systems of thought where they are not satisfied. For example, if concentrated sulphuric acid is added to water, a dilute solution results; but if water is added to concentrated sulphuric acid, there is a minor explosion, producing a fountain of sulphuric acid (Fig. 3.5). It is worth pointing out that the 'Laws' cannot be imposed on an existing system of operations—they are either satisfied or not. What we are saying is that ordinary addition and multiplication of integers is such that Laws 1–6 always hold.

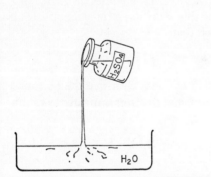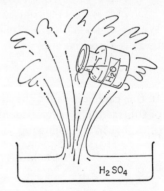

FIGURE 3.5

Since addition is initially defined as an operation on two integers, we add three of them by first adding two and adding the result to the third. Law 3 formalizes the result that it doesn't matter which pair is added first, and Law 4 does the same for multiplication. Law 5 gives the connection between the arithmetic operations. We can represent $4.(2+5)$ as in Fig. 3.6, which also represents 4.7.

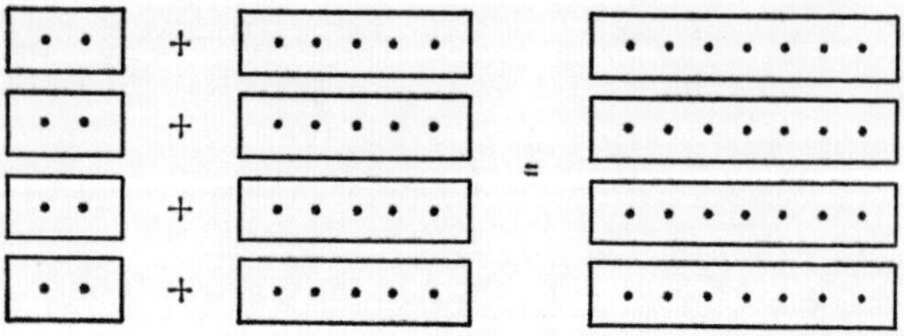

FIGURE 3.6

There are two other operations which are sometimes possible in N. The first of these is *subtraction*. If a, b are two whole numbers, and there is a third whole number, x, such that:

$$a + x = b$$

then we say that this whole number, x, is the result of subtracting a from b, and we write

$$x = b - a.$$

In this case we say that b is larger than a, and we write $b > a$.

Division is the other operation which can sometimes be carried out. We say that b is divisible by a, or that a divides b (in symbols $a|b$) if there is a positive integer, x, such that:

$$ax = b.$$

In this case, x is called the quotient b/a. It is clear that the operation of subtraction is inverse to addition, and division is inverse to multiplication, in the following sense:

$$(a + c) - c = a \qquad (b - c) + c = b$$

$$\frac{b}{a} \cdot a = b \qquad b \bigg/ \frac{b}{a} = a$$

Finally, it is convenient to introduce the integer *zero*, which we denote by 0, to be the number of elements in an empty set. Then, clearly we can define addition and multiplication so that:

$$a + 0 = a \quad \text{(for all } a \text{ in } N\text{)}$$
$$a \cdot 0 = 0 \quad \text{(for all } a \text{ in } N\text{)}$$

and

$$a - a = 0.$$

INTRODUCTION

Since very large collections have to be counted, mankind has introduced into his languages different methods of representing integers. When we say that the box on the right of Fig. 3.4 has twenty dots in it, we are using the decimal system. This uses ten digit symbols, 0, 1, 2, ..., 9, for zero and the first nine whole numbers, and expresses larger integers by adding multiples of powers of ten:

$$1, \quad 10, \quad 10.10, \quad 10.10.10 (= 10^3), \quad 10^4, \ldots$$

Thus twenty means $2.10 + 0$, and three hundred and seventy two means:

$$300 + 70 + 2 = 3.10^2 + 7.10 + 2$$

and is denoted shortly in the decimal system by the symbols 372. In fact if we consider the sequence $\{10^1, 10^2, \ldots, 10^n, \ldots\}$ of powers of 10, this will eventually get larger than any fixed integer z, and we can express z in the form:

$$z = a_n . 10^n + a_{n-1} . 10^{n-1} + a_{n-2} . 10^{n-2} + \ldots a_1 . 10 + a_0$$

where 10^n is the largest power of 10 that is not greater than z, and each of the $a_0, a_1, a_2, \ldots a_n$ is one of the digit symbols $0, 1, 2, \ldots 9$. We then represent z in the decimal system by $a_n a_{n-1} a_{n-2} \ldots a_1 a_0$. The decimal system using powers of ten is now universally accepted in all advanced languages, but it is worth pointing out that any whole number, other than 1, can be used instead of 'ten' as a base for a system of enumeration. Number 2 as a base has become important because only two digits, 0 and 1, are needed, which greatly simplifies computation on electronic computing machines. The disadvantage of base 2 representation is that a large number of places is needed for quite a small whole number:

$$\text{thirty two} = 1.2^5 + 0.2^4 + 0.2^3 + 0.2^2 + 0.2 + 0$$

so this would be represented in base 2 as 100 000.

Some mathematicians have gone to great lengths to give an abstract construction of the set N of whole numbers starting from simple axioms. We are going to assume that this work has been done, and that it agrees with our intuition. This means that we assume that whole numbers exist, and that arithmetical operations have been defined on them to agree with Laws 1–6.

Exercise 3a

1. Express 'thirty two' and 'one hundred and thirty six' in the systems with bases 5, 7, 12. What do the symbols 1111 and 313 mean in these systems?
2. Consider the problem of economically describing the whole numbers from zero to one thousand. If we are using base a, we need words for the digits $0, 1, 2, \ldots a-1$, and for various powers $a, a^2, a^3, \ldots$ up to the largest power not greater than one thousand. Which of the bases $a = 2, 3, 4, \ldots, 14$ requires the smallest number of words to represent all the numbers up to one thousand?

3.2 Ordering and well ordering

It is part of our intuitive picture of whole numbers that we can always compare them. That is, given any two different whole numbers, one of them is always larger than the other. Formally, if $a, b \in N$, then one and only one of the following is true:

$$a = b \quad \text{or} \quad a > b \quad \text{or} \quad b > a.$$

This defines a relation on the set $N \times N$. It clearly has the property that:

$$a > b, b > c \Rightarrow a > c$$

and it is clear that $a > 0$ for all a in N. The mathematician says that the set N is *totally ordered* by the relation '$>$'. Further, the order structure of N is preserved by the arithmetical operations in the following sense:

Law 7 $b > a \Rightarrow b + c > a + c$ for all $c \in N$
Law 8 $b > a \Rightarrow bc > ac$ for all $c \in N$

All the Laws 1–8 satisfied by the integers can be proved starting from a small number of simple assumptions. We shall not stop to do this as these Laws are very much part of our experience and we have no difficulty in believing them.

However, the fact that the system N of natural numbers is ordered turns out to be insufficient when we come to use N in mathematical arguments. The basic reason for this is that we shall often want to

ORDERING AND WELL ORDERING   43

prove that a result is true for every natural number $n \in N$. Since N is not a finite set it would take an unlimited amount of time to prove the result for each integer 1, 2, 3, . . . in turn. This means that we require a technique of proof that will enable us to deduce the validity of a result for every integer from a small number of steps that there is a hope of carrying out. This basic technique, which we discuss in Section 3.3, is called the 'principle of mathematical induction'.

Now the fact that proof by induction is valid is an additional property of the set N and the order relation $>$. This is equivalent to axiom I.

AXIOM I

The order of N is such that every non-empty subset $E \subset N$ has a smallest element.

This means that for each $E \neq \emptyset$, $E \subset N$ there is an integer $m \in E$ such that, for all $x \in E$ either $x = m$ or $x > m$. Any ordered set satisfying this condition is said to be *well ordered*. Now if D is any set which is totally ordered, this means that for any two elements a, b in E which are different we can always say which of these comes before the other. Clearly if we have such an ordering of the whole set D, we also have the same ordering for every subset $A \subset D$. We call the ordering a well ordering if each such subset A has a first element which comes before all others.

Example 1. If we consider the fractions of the form a/b with $a, b \in N$ these can be ordered by saying:

$$\frac{a}{b} < \frac{c}{d} \text{ if, and only if } a.d < b.c.$$

This gives the usual notion of ordering of fractional parts of a whole (*e.g.*, $\frac{2}{3} < \frac{3}{4}$). This is not a well ordered set (though it is totally ordered) for there is no smallest positive fraction. It is amusing that we can use our axiom to prove a theorem.

THEOREM

1 *is the smallest positive integer.*

Most people do not think it necessary to prove this theorem because they know that each natural number a satisfies either $a = 1$ or $a > 1$. What we actually do is to show that, if we assume that N satisfies Laws 1–8, and is totally ordered, and satisfies the well ordering axiom, then our theorem follows.

For suppose there is at least one positive integer c with $1 > c$. Then the set E of those positive integers c such that $1 > c$ is non-empty. By the well ordering axiom, E has a least member m, and $0 < m < 1$. But now we can multiply this inequality by m, use Law 8 and obtain $0 < m^2 < m < 1$, so that m^2 also belongs to E and is smaller than m. (For example, if $m = \frac{1}{2}$, then $0 < (\frac{1}{2})^2 < \frac{1}{2} < 1$.) This contradicts the fact that m is the least member of E. The assumption that the theorem is false leads to absurdity, so the theorem is proved.

3.3 Mathematical induction

The theorem proved in Section 3.2 is that there is no integer $m \in N$ such that $1 > m > 0$. It follows that, for any $a \in N$, there is no integer m such that $a + 1 > m > a$. Thus each positive integer a has an immediate successor $a + 1$, and (for $a > 1$) an immediate predecessor $a - 1$ in the ordering. This allows us to prove a general result by showing firstly that it is valid for $a = 1$, and secondly that its truth for a particular integer a implies its truth for the successor of a. Let us try to formulate this precisely.

THEOREM

Suppose that, for each $n \in N$, we have a statement H_n involving the integer n and:
 (i) *H_k is true for some fixed integer k:*
 (ii) *for each integer n such that $n > k$ or $n = k$, the truth of H_n implies the truth of H_{n+1}.*
Then it follows that H_n is true for every $n > k$.

For suppose that conditions (i) and (ii) are satisfied. Let E be the subset of N consisting of those integers $n > k$ for which H_n is false. If this set E is empty then H_n is true for every $n > k$ and our principle is established. However, if E is not empty, it has a least element m. Since $m > k$, we can consider $m - 1$ which must satisfy $m - 1 = k$ or $m - 1 > k$ since there are no integers between $m - 1$ and m. But $m - 1 \notin E$, so H_{m-1} is true. Now apply (ii) with $n = m - 1$, and we deduce that H_m is also true, which contradicts the fact that $m \in E$. This contradiction establishes the principle.

The above proof may be difficult to follow for the reader not experienced in mathematical arguments. It may help to think of a Venn diagram in which $S = N$, the set of positive integers, E is a subset containing some elements and m is the first element in E (Fig. 3.7). We use the particular element m to obtain a contradiction.

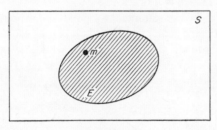

FIGURE 3.7

It is worth remarking that the principle of induction is not only useful for proving results, but is also needed in making definitions. For example, if $a \in N$, we know that a^2 is a short way of writing $a.a$; $a^3 = a.a.a$; $a^4 = a.a.a.a$; . . ., but what do we understand by a^n where n is an arbitrary integer? The principle of induction justifies a definition of a^n step-by-step. Thus, suppose for some integer k we have already defined a^k. Then we put:

$$a^{k+1} = a.a^k$$

and this extends our definition from k to $k + 1$. This procedure will therefore lead to the value of a^n for each integer n, since we can take, for H_n in the formulation of the principle of induction, the statement 'a^k has been defined for $k = n$'. Then obviously H_1 is true, and the truth of H_n implies the truth of H_{n+1}, so H_n must be true for all $n \in N$.

In the traditional school mathematics syllabus, the method of induction is used to find the sum of the first n integers:

$$1 + 2 + 3 + \ldots + n = \frac{n(n+1)}{2}$$

or the sum of a geometrical progression:

$$a + ar + ar^2 + \ldots ar^n = a\frac{1 - r^{n+1}}{1 - r}.$$

Instead of considering proofs of these results we illustrate the method by other examples.

Example 2. Prove that the sum of the interior angles in a convex polygon of $n + 2$ sides is n times $180°$.

Here we let H_n be the statement for a particular integer n. Then H_1 is true by elementary geometry, for it is just the result that the angles in any triangle add up to $180°$. Suppose we know that H_n is true. Let P_{n+3} be any convex polygon with $n + 3$ sides. Cut off a triangle by using three consecutive vertices of P_{n+3} (see Fig. 3.8). This

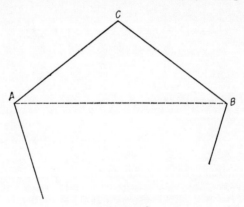

FIGURE 3.8

replaces two sides of P_{n+3} by a single side, thus forming a polygon P_{n+2} with $n + 2$ sides. If C is the vertex in P_{n+3} but not in P_{n+2} and A, B are the adjacent vertices, then the original angle at A is the sum of $C\hat{A}B$ and the new angle at A. A similar argument applies at B. The sum of the interior angles in P_{n+3} is therefore obtained by adding the angles in the triangle ABC to those in the new polygon P_{n+2} which we are assuming sum to $n.180$. Hence the angles in P_{n+3} sum to:

$$n.180 + 1.180 = (n+1).180$$

so that H_{n+1} is true. We have therefore proved that both conditions (*i*) and (*ii*) in the principle of induction are satisfied, so H_n is true for all $n \in N$.

Example 3. Prove that $2^n > n$ for each positive integer n.

Let H_n state that $2^n > n$. Clearly H_1 is true, since $2^1 (=2) > 1$. But, if H_n is true, then $2^{n+1} (=2.2^n) > 2n (=n+n) \geqslant n+1$, for all n, so H_{n+1} follows. By induction, H_{n+1} is true for all n.

Great care is needed in seeing that both conditions (*i*) and (*ii*) on page 44 are established in using the principle of induction. If we are sloppy in our thought, it is very easy to prove nonsense. Let us illustrate by giving a specious proof.

Example 4. The people present in any room all have the same age.

Let H_n be the statement that, if a room contains n persons, they are all of the same age. Since every room must contain a whole number of people it is sufficient to show that H_n is true for every $n \in N$. Now apply the principle of induction:

(*i*) H_1 is true, since if there is only one person in the room, he (or she) is the same age as himself.

(*ii*) Assume the truth of H_n and consider a fixed room containing $(n+1)$ persons. Number these inmates $1, 2, \ldots, n+1$ and first send person number 1 out of the room. The room now contains n people and by hypothesis they are all of the same age. Call number 1 back and send out person number $(n+1)$. The room again has n people in it—by hypothesis all of the same age. But now person number 1 is the same age as persons numbered 2 to n, and the same is true of person number $(n+1)$.

It follows that all $(n+1)$ persons are of the same age, and **we** have deduced the truth of H_{n+1}. (But see question 2 in Exercise 3b.)

Example 5. Suppose we have a set containing n elements: how many distinct subsets containing r elements are there?

The answer can be obtained by induction. The number of subsets is:

$$\binom{n}{r} = \frac{n!}{(n-r)!\, r!}$$

where $n!$ (read this as 'n factorial') stands for the product of the integers $1, 2, \ldots, n$.

Example 6. The method of induction can also be used to prove the binomial theorem.

For any positive integer n:

$$(a+b)^n = a^n + \binom{n}{1}a^{n-1}b + \ldots + \binom{n}{r}a^{n-r}b^r + \ldots + b^n$$

Here the ... indicates that we add the terms of the form $\binom{n}{r}a^{n-r}b^r$ for integers r between 1 and $(n-1)$ giving a total of $(n+1)$ terms in the expansion. (See pages 16 and 17 of Courant and Robbins, 'What is Mathematics?' for a detailed discussion of the binomial theorem.)

Exercise 3b

1. Prove by induction that, for any positive integer n, the sum of the squares of the first n integers can be given by:

$$1^2 + 2^2 + 3^2 + \ldots + n^2 = \frac{n(n+1)(2n+1)}{6}.$$

*2. Where is the flaw in the argument for Example 5? Hint: check that the condition (*ii*) of the principle of induction has been proved for every $n \geqslant 1$.

3.4 Divisibility properties

We saw in Section 3.1 that, for a given pair of integers $a, b \in N$ the equation $ax = b$ does not always have a solution x in N. If it has a solution, then it is unique, for suppose:

DIVISIBILITY PROPERTIES

then
so that
and
$$ax = b = ay$$
$$ax - ay = b - b = 0$$
$$a(x - y) = 0$$
$$x = y, \text{ since } a \neq 0.$$

When $ax = b$ has a solution in N we write a divides b or $a\,|\,b$. In general, it is important to know that we can always divide b by a to give a quotient q and a remainder r which is smaller than a. Let N_+ be the set of positive integers together with the number zero. We can formulate a theorem, known as the Euclidean algorithm.

EUCLIDEAN ALGORITHM

Given integers a, b in N there are unique integers q, r in N_+ with:

$$b = aq + r \qquad (0 \leqslant r < a)$$

We first prove that q, r are unique if they exist. Suppose q_1, r_1; q_2, r_2 both satisfy the relation, then $b = aq_1 + r_1 = aq_2 + r_2$. If $r_1 > r_2$, then $r_1 - r_2 = a(q_2 - q_1)$, so that $\dfrac{r_1 - r_2}{a} = q_2 - q_1$. But $q_2 - q_1$ is a whole number, so that $\dfrac{r_1 - r_2}{a}$ is a whole number. Hence $(r_1 - r_2)$ is divisible by a. But $(r_1 - r_2)$ is smaller than a, so that this is impossible, since 1 is the smallest positive integer, unless $r_1 = r_2$, $q_1 = q_2$.

Now let E be the set of integers in N_+ of the form $b - ax$. E is not empty, since $x = 0$ gives the integer b in N. By the well ordering axiom the set E has a smallest member r which corresponds to x equal to an integer q in N_+. Then certainly we have $b = aq + r$, and we only have to show that $a > r$.

Now, if $r \geqslant a$, then $b - a(q+1) = r - a$ would be a smaller integer in E, contradicting the definition of r. Hence the Euclidean algorithm is completely established.

The theorem we have just proved is another example of a result which we learnt to believe in by experience. Let us discuss one of the arithmetic results which can be deduced from the Euclidean algorithm, which is of fundamental importance in studying the structure of numbers.

Greatest common divisor

Given two integers a, b in N, we call d in N, the greatest common divisor (GCD) of a and b if d is a divisor of both a and b, and any x in N which divides both a and b also divides d. Thus $d\,|\,a$, $d\,|\,b$ and $x \in N$, $x\,|\,a$, $x\,|\,b \Rightarrow x\,|\,d$.

Note that the word 'greatest' here really means that d is a multiple of any other divisor.

THEOREM

Any two positive integers a, b have a unique GCD *in N, denoted by (a, b). There are integers u, v (not unique, and not both positive) such that $(a, b) = ua + vb$.*

[Here if u is negative this is of the form $vb - ka$, with v, $k \in N$, whereas if v is negative it is of the form $ua - kb$, with u, $k \in N$. Note that we are using the notation of an ordered pair (a, b) to stand for the unique integer d which is the GCD of a and b. The convenience of this rather bizarre notation will become apparent in the proof.]

This theorem is proved by a repeated application of the Euclidean algorithm. Dividing b by a gives $b = aq_1 + r_1$ $(0 \leqslant r_1 < a)$. Now if $x \in N$ divides both a and b it must divide r_1, and similarly if it divides a and r_1 it must divide b. It follows that $(a, b) = (a, r_1)$. If $r_1 \neq 0$, repeat the argument on a and r_1, giving $a = r_1 q_2 + r_2$ $(0 \leqslant r_2 < r_1)$. If $r_2 \neq 0$, divide r_1 by r_2, and continue until ultimately there is a zero remainder. This must happen in a finite number of steps, since the remainder decreases by at least 1 each time.

Suppose n is an integer such that the last two steps are:

$$r_{n-2} = r_{n-1} q_n + r_n \qquad (0 \leqslant r_n < r_{n-1})$$
$$r_{n-1} = r_n q_{n+1}.$$

Since r_n divides r_{n-1} we have $(r_n, r_{n-1}) = r_n$. Now as we saw for the first step, each of the pairs a, b; a, r_1; r_1, r_2; ...; r_{n-1}, r_n have the same set of common divisors. Hence:

$$(a, b) = (r_{n-1}, r_n) = r_n$$

and the GCD must be the last non-zero remainder in this (finite) process.

PRIME NUMBERS

The uniqueness for the GCD follows from the fact that, if d_1, d_2 both satisfy the conditions for a GCD then $d_1 | d_2$ since d_1 is a common divisor and d_2 is a GCD. Similarly $d_2 | d_1$. This implies $d_1 = d_2$. The evaluation of integers u, v such that $(a, b) = ua + vb$ can also be carried out from the system of equations, by successive substitution giving each r_i in turn as a sum of the form $u_i a + v_i b$:

$$r_1 = b - aq_1 = 1.b + (-q_1).a$$
$$r_2 = a - q_2 r_1 = (-q_2)b + (1 + q_1 q_2)a$$

and so on, by induction.

We can also find the GCD of two integers by expressing both integers as a product of primes. The fact that this method works can, however, only be justified by proving something like the above theorem. We will discuss prime factorization in Section 3.5.

Exercise 3c

Use the Euclidean algorithm as in the proof of the last theorem to find the GCD of the following pairs of integers:

(315, 385)   (144, 1000)   (1001, 7655)

In each case find integers u, v such that:

$$(a, b) = ua + vb.$$

3.5 Prime numbers

A prime number is any integer p in N other than 1, such that the only integers in N which divide p are 1 and p.

Everyone is familiar with the first few primes:

2, 3, 5, 7, 11, 13, 17, 19, 23, 29, 31, 37, 41, ...

Tables have been prepared listing all the primes up to 10^7, and some very much larger numbers are known to be primes. The prime numbers have fascinated man ever since he started to think about numbers. We gave the proof in Section 1.7 of Euclid's theorem that there are infinitely many primes.

It is somewhat surprising that no one has yet been able to decide whether there are infinitely many primes p for which $p + 2$ is also a prime. This is a famous unsolved problem of number theory which is striking because it can be stated so simply.

There are other questions to be asked about the sequence of primes $p_1, p_2, p_3, \ldots, p_n, \ldots$ written down in increasing order of magnitude. If you look at this sequence by examining a table of primes, you notice:

(*i*) the sequence looks irregular if you only consider a small part of it—like the primes between 5010 and 5080, or between 8010 and 8080;

(*ii*) prime numbers become rarer on average among the larger integers and one feels that if an arbitrary integer greater than, say, 10 000 000 is considered, its chance of being a prime is very small;

(*iii*) if you count the number of primes in large blocks, say between 2^k and 2^{k+1} for $k = 5, 6, \ldots$, one obtains a sequence of integers which is fairly regular and does not seem to tend to zero.

Since the sequence of primes is completely determined (though it is not possible to give a simple arithmetical formula which yields only primes, let alone gives all the primes) we can define a function $f: N \to N$ by:

$$f(n) = \text{number of primes } p \leqslant n.$$

For example, $f(10) = 4$, $f(40) = 12$, $f(100) = 25$. Since there are infinitely many primes we know that $f(n)$ grows arbitrarily large as n grows, but our observation (*ii*) above would lead us to suspect that $f(n)$ is small compared with n for large values of n. The proportion of primes among the first n integers is defined to be $d(n)$, where:

$$d(n) = \frac{f(n)}{n}$$

and this can be computed for particular values of n. It is possible to prove, without too much machinery, that simply by making the integer n large enough, we can arrange for this proportion $d(n)$ to be as small as we like.

The mathematician Gauss, at the beginning of the nineteenth century, noticed that $d(n)$ in fact gets small in a very regular manner for large values of n. We shall discuss this as Example 5 in Chapter 9 (p. 160), when we have considered the idea of convergence.

PRIME NUMBERS

The theory of numbers has been very important in the development of mathematical thought. Its beauty lies in the simplicity of the problems it poses. We have mentioned some of the results related to the distribution of primes, but there are many other fascinating problems concerning the structure of N, the set of whole numbers. The reader who is interested should read a book introducing number theory, such as Davenport's 'The Higher Arithmetic'.

However, let us return to the task of proving some facts about primes which we learnt by experience and have always believed. Even though the proofs are not very difficult, they do need to be carried out, for the fact of our experience, that the result is true in every case we have examined, does not constitute a proof that it is always true.

If p is a prime, $a, b \in N$ and $p \mid ab$ then $p \mid a$ or $p \mid b$. For suppose $p \mid ab$, but p is not a divisor of b. Then the only divisor of both p and b is 1, and so $(p, b) = 1$. Using the theorem of the last section, there are suitable integers u, v such that:

$$1 = up + vb.$$

Multiplying through by a gives:

$$a = upa + vab,$$

and if $p \mid ab$ it is clear that the right-hand side is divisible by p so that $p \mid a$.

Now, if $n \in N$ and n is not prime, it has divisors other than 1 and itself. The smallest such divisor (which exists by the well ordering axiom) must be a prime. Hence any n which is not a prime is divisible by a prime. By dividing out this prime divisor and repeating the argument we can express any integer $n > 1$ as a product of a finite number of primes. We can now prove our main result.

THEOREM

The expression of an integer $n > 1$ as a product of primes is unique apart from the order of the factors.

Suppose we have two factorizations into primes:

$$n = p_1 p_2 \cdots p_n = q_1 q_2 \cdots q_m$$

Then, since $p_1|n$, we have $p_1|q_1(q_2 \ldots q_m)$ so that either $p_1|q_1$ or $p_1|q_2 (q_3 \ldots q_m)$. Repeat the argument so that either $p_1|q_1$ or $p_1|q_2$ or $p_1|q_3(q_4 \ldots q_m)$, and so on. By induction, there must be some suffix i, such that $p_1|q_i$. As these are both primes, we must have $p_1 = q_i$ and this term can be cancelled from the product leaving:

$$p_2 p_3 \ldots p_m = q_2' q_3' \ldots q_m$$

where the right-hand side consists of the remaining factors q_j. This argument can now be repeated with each of $p_2, p_3, \ldots, p_n$ in turn, showing that each of the prime numbers $p_1, \ldots p_n$ occurs among the primes q_i. After they have all been cancelled there cannot be any factors q_i left. Hence $n = m$, and the factors q_i are just a rearrangement of the factors p_i.

Now that we have proved that prime factorization of integers is unique, it is legitimate to use it as a method of finding the GCD of two integers. Thus the common arithmetical techniques can be justified by using the basic properties of the integers which we have formulated. This is gratifying for most of us who have found these techniques very useful in practice.

Exercise 3d

1. The result $p|ab \Rightarrow p|a$ or $p|b$ is true for primes p, but it is not true if p is replaced by an integer n which is not a prime. Try $p = 6$, $a = 15$, $b = 16$.
2. How many divisors has 144?
3. If we factorize an integer n so that:

$$n = p_1^{\alpha_1} p_2^{\alpha_2} \ldots p_r^{\alpha_r}$$

where the p_i are distinct primes each raised to an integer power α_i, write down a formula for the number of integers k which divide n exactly.

3.6 Goldbach's conjecture (1742)

The conjecture can be stated simply:

Every even number, other than 2, can be expressed as the sum of two primes.

The beauty of the conjecture lies in its simplicity; nevertheless, it has not yet been settled, though there is now more hope of a solution in a finite time. A Russian mathematician, Vinogradoff, has shown that all sufficiently large integers can be expressed as the sum of at most 4 primes, but this is still quite a long way from Goldbach's conjecture. There is no even number known that cannot be expressed as the sum of two primes, but this does not prove the conjecture either—although it does mean that there is no known examples to disprove the conjecture.

4 WHAT IS A REAL NUMBER?

4.1 Introduction

To answer the question in our chapter title properly would require a complete book substantially longer than this one. Yet the question is so basic that we must not shirk the need to have some measure of understanding of it. There are several distinct ways of giving a logically exact definition of the set of real numbers: we shall not describe any one of these systematically, but rather examine in this chapter the intuitive concepts which lie behind the problem. Our object will be to build up an adequate picture in our mind of the properties which we expect of the set R of real numbers—properties which can be proved to hold by a careful look at the foundations of the definition of the set R.

The reader may wonder why the word 'real' is used as an adjective describing numbers. This is an example of a word which already has a meaning in the English language being taken over by the mathematician and given a precise technical meaning. A particular set of 'numbers' with the properties we are going to describe is called the set of 'real numbers'. In other contexts the mathematician invents a new word (like 'subset') to describe a notion precisely and this word is then added to his vocabulary. However, it is more usual to adapt an existing word (like 'real') and give it a new precise meaning.

There are two main avenues by which we can approach the subject of real numbers. Both of these start from the set N of natural numbers or positive integers which we discussed in Chapter 3. The first method of attack is to enlarge the set N step by step so that the arithmetical operations of subtraction and division can always be carried out, and the results of carrying out the operations satisfy simple algebraic laws (Laws 1-6 in Section 3.1, together with some additional ones). This procedure leads first to the addition of 0, then to

INTRODUCTION 57

the set of all integers (including negative integers), and then to the rational numbers of the form a/b where a, b are integers, and $b \neq 0$. We shall give an intuitive description of this procedure in Section 4.2, and then notice that we still do not have a large enough set to solve equations like $x^2 = 2$.

An alternative approach is to consider the problem of measuring physical quantities, such as length, area, weight and time. For example, let us try to find the weight of a given mass of lead. We choose a unit of measurement, say the kilogramme, and count how many of these units are needed to produce the same weight as the lump of lead. We would not expect to get an exact answer like 34 kg —it is much more likely that we would discover that the weight of the lump of lead is more than 33 kg but less than 34 kg (Fig. 4.1).

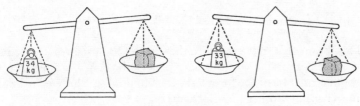

FIGURE 4.1

When this happens we divide our original unit into a whole number n of equal parts. In real life these sub-units are given different names —a metre is divided into 100 centimetres, a kilogramme is divided into 1000 grammes, an hour is divided into 60 minutes, a litre is divided into 1000 millilitres, and so on. If we now measure the weight of lead in grammes, we might find that 33 128 grammes are not enough, and 33 129 grammes are too much. Thinking of a gramme as one-thousandth of a kilogramme, the lead weighs between $\frac{33128}{1000}$ and $\frac{33129}{1000}$ kg. In practice we could go on to subdivide the unit further to obtain a more accurate measurement of the weight using fractions of a kilogramme with a larger denominator.

It is intuitively clear that we can measure a physical quantity, like weight, as accurately as we please in terms of a prescribed unit, such as one kilogramme, by using fractions of the form a/b, where $a, b \in N$. At first sight one might expect to be able to measure exactly using a fraction with a sufficiently large denominator, *i.e.*, by dividing

c

the original unit into a sufficiently large integer number of equal pieces. The fact that this is not possible was known to the Greek school of mathematicians at the time of Pythagoras. Consider a right-angled triangle OAB, in which the sides OA, AB about the right angle each have length 1 metre (Fig. 4.2). Then:

$$OB^2 = OA^2 + AB^2 = 2.$$

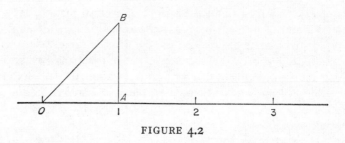

FIGURE 4.2

This means that, if we measure the length OB in metres or fractions of a metre, the square of this length has to be 2. We can see that the length cannot be of the form m/n. For suppose m, n have no common integer factors (other than 1), and:

$$\left(\frac{m}{n}\right)^2 = 2.$$

(If there were any common factor we could divide both m and n by it —giving an equal fraction with a smaller denominator.) Now multiply by n^2 to give:

$$m^2 = 2n^2.$$

The prime $2 \mid 2n^2$, (read thus: the integer 2 divides the integer $2n^2$ exactly) so $2 \mid m^2$. That is, $2 \mid m.m$, so $2 \mid m$. Put $m = 2k$ for a suitable $k \in N$. Then:

$$(2k)^2 = 4k^2 = 2n^2$$
$$2k^2 = n^2.$$

and a repetition of the argument shows that $2 \mid n$. This means that m, n have the common factor 2, contradicting the assumption that they have no common factor.

Thus the problems of measurement lead to the same kind of

RATIONAL NUMBERS   59

difficulties as the problem of examining the algebraic structure to see whether it is possible to solve equations like $x^2 = 2$ or $x^5 = 4$.

Exercise 4a

1. Show that there is no number x of the form m/n such that $x^3 = 7$.
*2. If $q \in N$, and q is not the square of any positive integer, show that it cannot be the square of any fraction: that is, there is no number $x = m/n$ such that $x^2 = q$. (Hint: consider one of the prime factors of q and show that it must divide both m and n.)

4.2 Rational numbers

In Chapter 3 we saw that the arithmetic operations of addition and multiplication satisfy:

Law 1 $\qquad a + b = b + a$
Law 2 $\qquad ab = ba$
Law 3 $\qquad (a+b) + c = a + (b+c)$
Law 4 $\qquad (ab)c = a(bc)$
Law 5 $\qquad a(a+c) = ab + ac$
Law 6 $\qquad 1a = a \qquad$ for all a
Law 7 $\qquad a + 0 = 0 + a = a \qquad$ for all a
$\qquad \qquad 0.a = 0 \qquad$ for all a

when applied to the set N_+ consisting of the positive integers together with zero. The disadvantage we noted was that we could not always subtract, and we could not always divide in the set N_+. The set of rational numbers, which we call Q, is the smallest set containing N_+ in which subtraction and division (except by zero) can always be carried out. This is ensured if we assume Law 8:

Law 8 $\qquad$ For every a there is a unique element $-a$ such that:
$$a + (-a) = -a + a = 0$$

Law 9 $\qquad$ For every $a \neq 0$ there is a unique element a^{-1} such that:
$$a.a^{-1} = a^{-1}.a = 1$$

for then we can define:
$$a - b = a + (-b)$$
$$\frac{a}{b} = a.b^{-1}$$

It is worth noticing that our assumptions imply that $-(-a) = a$, since $-(-a)$ is the unique element satisfying $-a + -(-a) = 0$ and we already know that $-a + a = 0$. Further, if we apply Law 5 to $1 - 1 = 0$ we obtain:

$$a(1-1) = 0$$

or $\quad a.1 + a(-1) = 0$

or $\quad a + (-1)a = 0 \qquad$ (by law 2)

so that $\quad (-1)a = -a \qquad$ (for every a)

Now using Law 4, we see that the familiar 'law of signs' is forced on us, namely that the product of two negative numbers is positive. For, if $a, b \in N$, then:

$$\begin{aligned}(-a)(-b) &= (-1)a.(-1)b \\ &= (-1).(-1).ab \\ &= -(-1)a.b \\ &= 1.a.b = a.b\end{aligned}$$

Note that we have not really proved anything. What we have done is to observe that if the set $Q \supset N_+$ exists and satisfies Laws 1–9, then the familiar rules for manipulation follow. Mathematicians have given procedures for constructing such a set Q using the properties of N which we discussed in Chapter 3. The details of these constructions are of interest and are not too difficult; but they are much too extensive to give here. The interested reader should consult an account such as that given by Roberts in 'The Real Number System in an Algebraic Setting'. For our purposes it is sufficient to note that the construction can be carried out, and that the Laws 1–9 give us the rules of manipulation to which we are accustomed.

It is possible to represent the elements of Q by fractions:

$$\frac{r}{s}, \text{ which we can also write as } r/s,$$

where $s \in N$ and either $r \in N_+$ or $-r \in N$, that is, s is a positive integer and r is an integer (positive, negative or zero). Such a representation is not unique for, if $a \in N$:

$$\frac{a\,r}{a\,s} \text{ and } \frac{r}{s}$$

must represent the same element of Q (note that if $a \in N$, then $a \neq 0$).

RATIONAL NUMBERS

Out of the many representations of an element x in Q, we can pick a 'preferred one' by saying that r, s must have no common factors. We can divide out any common factors which occur to give a representation of each element in Q by a fraction r/s, for which there is no positive integer a such that $r = at, s = au$ with t, u integers. In common language, r/s is then said to be 'in its lowest terms'. Such a representation can be shown to be unique.

It helps our intuition to have a geometrical interpretation of the set Z of all integers, and the set Q of rational numbers. On a straight line —usually thought of as horizontal—mark a point 0 to represent zero and a second point to represent 1 to the right of 0. The positive and negative integers are then represented as a set of equidistant points, the positive numbers to the right in their natural order, and the negative integers to the left in reverse order (see Fig. 4.3).

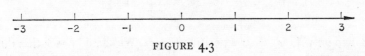

FIGURE 4.3

If we use the distance from 0 to 1 (which is the same as the distance between any pair $a, a+1$ of adjacent points on the line) as a unit for measuring length, then the distance from 0 to a will be a for $a \in N$, and it will be $-a$, if a is negative. Now consider a fraction r/s as a point of Q. We know that $s/1 = s$ and that $\frac{1}{s} \cdot s = 1$, so if we were to take $\frac{1}{s}$ as a unit of length on the line, then s of these mini-units would be equal to the original unit. This is the clue—divide the segment from 0 to 1 into s equal parts and the first division point will represent the number $1/s$. If $r > 0$, then r/s will be represented by a point to the right of 0 whose distance from 0 is r of these mini-units, while if $r < 0$, then it is represented by a point to the left of 0 in the number line. The effect of this operation is to put a lot of new points on the

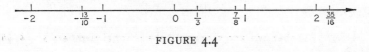

FIGURE 4.4

number line between each of the points representing an integer (see Fig. 4.4). It is easy to check that, if we carry out our operation with two fractions $\frac{au}{av}, \frac{u}{v}$ which represent the same rational number (in common language we say these numbers are equal), then, since av units of length $1/av$ have the same length (namely, 1 unit) as v units of length $1/v$, the point on the number line which represents a/av must be the same as that for $1/v$. It follows that the same point on the line represents both au/av and u/v. This means that two different fractions representing the same rational number will correspond to a single point on the number line, and different elements of Q are represented by distinct points on the number line.

We saw in Chapter 3 that N was an ordered set, and that the ordering was related to the arithmetical operations (see page 42). We can extend this order to the set Q by saying that:

$$x = \frac{a}{b} > \frac{c}{d} = y$$

if, and only if, $ad - bc$ is a positive integer. We should check that this definition is proper; that is, that it makes no difference which of the possible fractions are used to represent the rational numbers x and y.

On the number line $x = a/b > c/d = y$ means that the point representing $x = ad/bd$ comes to the right of the point $bc/bd = y$. The ordering of Q is therefore faithfully represented by our intuitive notion of order on the line.

It is also important to note that the rational numbers Q with the ordering $>$ satisfy:

Law 10 $a > b \Rightarrow x + a > x + b$
Law 11 $x > 0$ and $a > b \Rightarrow xa > xb$

Mathematicians call any set F in which there are operations of addition and multiplication defined and satisfying Laws 1–9 a *field*. If F is totally ordered and the order relation satisfies Laws 10 and 11 then we say that F is an *ordered field*.

Exercise 4b

Draw a number line and mark on it the rational numbers $\frac{1}{5}, -\frac{2}{5}, \frac{14}{25}, \frac{12}{8}, \frac{3}{2}, \frac{24}{16}$.

Example 1. It is worth noticing that it is possible for a set containing only a finite number of elements to be a field. For example, consider $F = \{0, 1, 2\}$, in which we define addition and multiplication by:

$$0+1 = 1 = 1+0 \qquad 0.1 = 0 = 1.0$$
$$0+2 = 2 = 2+0 \qquad 0.2 = 0 = 2.0$$
$$1+2 = 0 = 2+1 \qquad 2.1 = 2 = 1.2$$
$$0+0 = 0 \qquad\qquad 0.0 = 0$$
$$1+1 = 2 \qquad\qquad 1.1 = 1$$
$$2+2 = 1 \qquad\qquad 2.2 = 1$$

It is easy to see that Laws 1–9 are satisfied by this set of three elements. However, there is no way of defining an order relation in F to satisfy Laws 10 and 11—for, if we adopt the obvious $2>1>0$, then $2+1 \not> 1+1$. Thus F is a field but not an ordered field. The same kind of construction can be used to define a field containing p elements, for any prime p. It is not possible to produce fields containing n elements for any integer n which is divisible by more than one prime.

Example 2. Every positive rational number can be represented as the sum of an integer in N_+ and a fraction a/b in which $b > a \geqslant 0$.

This may seem obvious to us, for we learnt how to do it at an early age. However, to prove that it is possible we can use the Euclidean algorithm (see Section 3.4) to give for any $x = \dfrac{r}{s} > 0$:

$$r = qs + a \qquad (0 \leqslant a < s)$$

so that
$$x = \frac{r}{s} = q + \frac{a}{s}.$$

The value of q (the integer part of x) does not depend on the representation $\dfrac{r}{s}$ of x for, if $x = \dfrac{tr}{ts}$, then:

$$tr = q(ts) + ta$$

and
$$\frac{tr}{ts} = q + \frac{ta}{ts}.$$

This means that, if we consider the measurement of weight in tons, hundredweights, quarters, pounds and ounces, where the ton is the basic unit, a weight of 253 quarters is:

$$\tfrac{253}{80} \text{ tons} = (3 + \tfrac{13}{80}) \text{ tons.}$$

4.3 Decimals

We saw in Section 3.1 that it has become common practice in all civilized languages to use the decimal representation for positive integers. The metric system of weights and measures extends this convention further and is more convenient for calculation than the complicated Imperial system involving many different subdivisions (as in Example 2 at the end of Section 4.2). The decimal representation has to be extended indefinitely if it is to work for every rational in Q. Example 2 shows that any positive rational number is the sum of an integer in N_+ and a positive rational which is smaller than 1. The integer has a representation in decimal notation (in fact we do not normally think of it except in this representation, for it is part of our language). It is therefore sufficient to deal with fractions less than 1.

A decimal fraction is a fraction which can be represented in the form:

$$\frac{a}{10^n}$$

for some integer n; for example, $\frac{1}{2} = \frac{5}{10}$, $\frac{1}{8} = \frac{125}{1000}$. Because the numerator of such a fraction can always be written in the decimal notation, we can split it up so that we have a sum of fractions with denominators 10^k, $k \leqslant n$, and each of the numerators is one of the digits 0, 1, 2, ..., 9:

$$\frac{125}{1000} = \frac{100}{1000} + \frac{20}{1000} + \frac{5}{1000} = \frac{1}{10} + \frac{2}{100} + \frac{5}{1000}.$$

We then choose to simplify the writing by putting this in shorthand:

$$\frac{1}{8} = 0 \cdot 125$$

where the place position after the 'decimal point' indicates the corresponding power of 10 in the denominator. For ease of reading, it is usual to put a zero in front of the decimal point when the number has no 'whole number' part. This alerts us to look for the decimal point which we might otherwise miss.

This means that every finite decimal:

$$0 \cdot a_1 a_2 \ldots a_n$$

where the a_i are integers, $0 \leqslant a_i \leqslant 9$, is a representation of the rational number:

$$\frac{a_1 a_2 \ldots a_n}{10^n} = \frac{a_1}{10} + \frac{a_2}{10^2} + \cdots + \frac{a_n}{10^n}.$$

DECIMALS

An immediate question is whether the reverse is true. Can every rational number be represented as a decimal? If we restrict ourselves to 'finite' decimals, the answer is clearly 'no'. It is sufficient to show that one particular rational number, say $\frac{1}{3}$, has no such representation. Suppose, for some $b, n \in N$:

$$\tfrac{1}{3} = \frac{b}{10^n}.$$

Then $\quad\quad\quad 10^n = 3b$
or $\quad\quad\quad 2^n . 5^n = 3b.$

But now the prime number 3 divides the right-hand side, but does not occur as a factor of the left hand side—and this contradicts the main theorem of Section 3.5. Let us illustrate this on the number line.

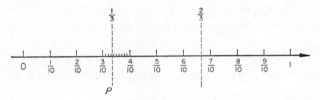

FIGURE 4.5

The point P on the line which represents $\frac{1}{3}$ lies between 0 and 1. If we divide the unit into 10 equal pieces each of length $\frac{1}{10}$, then P lies in the fourth such interval (see Fig. 4.5). That is, giving symbols to the ends of this interval:

$$x_1 = \tfrac{3}{10} < \tfrac{1}{3} < \tfrac{4}{10} = y_1$$

The piece of the line from x_1 to y_1 can now be divided into 10 equal pieces each of length $\frac{1}{100}$, and again P does not lie at a division point but in the fourth such little piece. That is:

$$x_2 = \frac{3}{10} + \frac{3}{10^2} < \frac{1}{3} < \frac{3}{10} + \frac{4}{10^2} = y_2.$$

The process can be repeated indefinitely giving a sequence of numbers:

$$x_n = \frac{3}{10} + \frac{3}{10^2} + \ldots + \frac{3}{10^n} < \frac{1}{3} < x_n + \frac{1}{10^n} = y_n$$

66   WHAT IS A REAL NUMBER?

This means that the rational number $\frac{1}{3}$ is greater than 0·33 ... 33, but less than 0·33 ... 34, where the number of digits may be arbitrarily large. We write:

$$\tfrac{1}{3} = 0·333 \ldots \quad \text{or} \quad \tfrac{1}{3} = ·\dot{3}$$

to indicate that we have to keep repeating the digit 3 indefinitely in order to express the rational $\frac{1}{3}$ in decimal notation. In fact all we have done is to justify the usual procedure of division:

$$3)\overline{1·0000 \ldots} \\ 0·333 \ldots$$

in which we pick the largest digit giving a result which is too small and then bring down the next digit.

This same sequence of operations can be done for any point P on the number line. For example, let us carry out the construction of Fig. 4.2, and try to measure the distance OB. The usual arithmetic procedure for finding the square root of a number has the effect of successively subdividing intervals of length 1, $\dfrac{1}{10}$, $\dfrac{1}{10^2}$, ... to obtain better and better approximations to the length of OP (see Fig. 4.6).

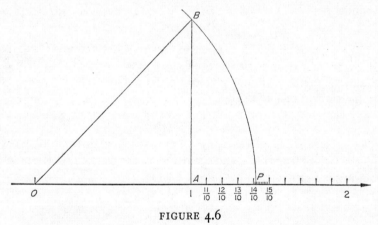

FIGURE 4.6

The digits in the decimal representation of P are derived from:

$$\begin{array}{llll}
1^2 & = 1 & <2 & < 2^2 & = 4 \\
(1·4)^2 & = 1·96 & <2 & <(1·5)^2 & = 2·25 \\
(1·41)^2 & = 1·9881 & <2 & <(1·42)^2 & = 2·0264
\end{array}$$

$(1·414)^2 = 1·999\ 396 < 2 < (1·415)^2 = 2·002\ 225$
$(1·4142)^2 = 1·999\ 961\ 64 < 2 < (1·4143)^2 = 2·000\ 244\ 49$

This means that, although P does *not* represent a rational number, we can find an infinite decimal:

$$z·a_1 a_2 \ldots a_n \ldots$$

where z stands for the whole number part of the representation, so that if we consider the finite decimal (which is also a rational number):

$$x_n = z + \frac{a_1}{10} + \frac{a_2}{10^2} + \ldots + \frac{a_n}{10^n},$$

then P lies on the line between the representations of x_n and $x_n + \frac{1}{10^n}$.

Now if we think of a finite decimal as being the same as the infinite decimal obtained by adding a string of digits, each zero, then all finite decimal representations are also infinite decimals. Thus:

$$\tfrac{1}{4} = 0·25 = 0·250\ 000 \ldots$$
$$\tfrac{1}{8} = 0·125 = 0·125\ 000\ 0 \ldots$$

This means that every rational number has a representation as an infinite decimal. Our argument for the point P, such that $OP^2 = 2$, would apply to any point P on the number line. For any point P on the line we can find an infinite decimal representation:

$$z·a_1 a_2 \ldots a_n \ldots$$

and this includes some points which do not represent rationals.

Is the converse true? Given an infinite decimal $z·a_1 a_2 \ldots a_n$ does this determine a unique point P on the number line? We know that, if such a point P exists then it lies between:

$$x_n = z + \frac{a_1}{10} + \frac{a_2}{10^2} + \ldots + \frac{a_n}{10^n} \text{ and } x_n + \frac{1}{10^n} = y_n.$$

These segments of the line, which we denote by $[x_n, y_n]$, have the property that each is inside the previous:

$$[x_{n+1}, y_{n+1}] \subset [x_n, y_n]$$

and $y_n - x_n = \frac{1}{10^n}$ grows smaller and smaller as n increases. It is a plausible assumption that there must be a unique point P in such a 'nest' of segments (see Fig. 4.7, which indicates why we call it a nest). Yet we know that, if the line contained only representatives of the set Q of rationals, our intuition is false. For there is an infinite decimal which represents $\sqrt{2}$, which we have proved is not rational. This means that, in order to satisfy our intuition that there are no 'holes' or 'gaps' in the number line we require a bigger set than Q. So we have to enlarge our number system further.

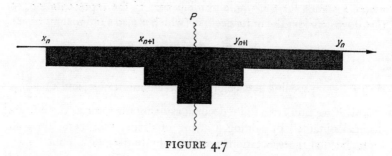

FIGURE 4.7

Example 2. It is possible to decide which infinite decimals represent rational numbers, though the proof of this requires more technique than we have available. A decimal is said to be 'recurring' if there is a block of digits $b_1 b_2 \ldots b_k$ which occurs again and again in the same order:

$$z = 0 \cdot a_1 a_2 \ldots a_n b_1 b_2 \ldots b_k b_1 b_2 \ldots b_k b_1 \ldots$$

is written:

$$z = 0 \cdot a_1 a_2 \ldots a_n \dot{b}_1 b_2 \ldots \dot{b}_k.$$

The integer k is called the period of the decimal. For example $\frac{1}{7} = 0 \cdot \dot{1}4285\dot{7}$ has period 6. An *infinite decimal* represents a rational number if, and only if, it is recurring.

Example 3. All our arguments using the base 10 are valid to any integer base. For base b, the digits a_1 satisfy $0 \leqslant a_i < b$ and we have an expansion of any rational:

$$\frac{m}{n} = z \cdot a_1 a_2 \ldots a_n \ldots$$
$$= z + \frac{a_1}{b} + \frac{a_1}{b^2} + \ldots + \frac{a_n}{b^n} + \ldots$$

DECIMALS

FIGURE 4.8

Exercise 4b

1. Represent $\frac{1}{11}, \frac{1}{13}, \frac{2}{13}, \frac{1}{17}, \frac{2}{17}$, as decimals and find the period in each case.
2. Assuming there is a number x such that $x^3 = 2$, find the first two places in the decimal representation of x.

4.4 System of real numbers R

Our discussion has illustrated the different stages in the development of the real number system (see Fig. 4.8). At each stage more precise concepts were introduced to meet a specific need. We have now realized that the set Q of rationals does not contain enough elements to allow us to solve simple equations like $x^2 = 2$. This means we have to take a bigger set. In our representation by decimals we have seen that every infinite decimal will correspond to a point if decreasing nests of segments with end-points in Q always contain a common point. That is, we would like to arrange that if:

$$a_1 \leqslant a_2 \leqslant \ldots \leqslant a_n \leqslant \ldots \leqslant b_n \leqslant \ldots \leqslant b_2 \leqslant b_1$$

and each of the $a_i, b_i \in Q$, then there is a 'number' x which is in all the segments:

$$a_i \leqslant x \leqslant b_i \qquad (i = 1, 2, \ldots).$$

It is easier to illustrate the situation by picking the segments out of the number line, and stacking them (see Fig. 4.9) to form 'a nest'.

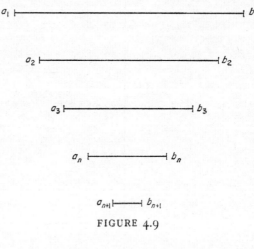

FIGURE 4.9

Now we want to preserve the important property of Q that it is an ordered field containing N. There are various ways of 'filling up' the holes in Q to make it continuous. One method is to take a point for each nest of intervals such that $b_n - a_n$ becomes arbitrarily small as n gets large, and then to identify those nests which must give the same point. However, the details are tedious, so let us just say that the extension can be done and the result is an ordered field R which contains Q and has the additional property that nests of closed segments of R always contain a point of R. This additional condition is called *completeness*. Thus addition and multiplication are defined in R and satisfy Laws 1–9 (see Section 4.2) and there is an order relation which satisfies Laws 10 and 11.

It can also be shown that each number in R has a representation as an infinite decimal:

$$z \cdot a_1 a_2 \ldots a_n \ldots$$

and that the representation is unique except for those numbers which come at one of the decimal division points. Thus, for example:

$$0 \cdot 459\ 999 \ldots = 0 \cdot 460\ 00 \ldots$$

Further, each infinite decimal now corresponds to a real number because of the completeness property. This means that we can think of R as the set of all infinite decimals which do not end in 9 recurring.

The rules of arithmetic tell us how to compute (that is add, subtract, multiply, divide) with finite decimals. Since we can approximate to real numbers by finite decimals, we can therefore calculate to any desired degree of accuracy the sum or product of two real numbers. We can therefore use our concrete representation as an infinite decimal to help us to find the result of arithmetical operations in R.

The reader who wants to understand thoroughly the construction of the set R will need to spend quite a bit of time on the problem. There is a rigorous account in Landau's 'Foundations of Analysis'.

4.5 Some properties of R

The development of the concept of real numbers has been one of the avenues of thought which has had the utmost importance in the deeper understanding of the nature of mathematics. Up to about the middle of the nineteenth century there was no precise logical basis for the set R. The calculus as developed by Newton and Leibnitz (and as taught in most secondary schools) does not examine the nature of the set R, although calculus can only be justified logically by a clear understanding of most of the properties of R.

In the 80 years from about 1840 to 1920 a great deal of research was done by mathematicians into this basic problem, so that to-day we are not only satisfied that the set R can be defined, but we also know a great deal about its properties. It is normal for students to do a one-year course at university level studying Analysis, the object of which is to understand the real numbers, and how they can be used to justify not only the Calculus, but also some more advanced tools requiring these concepts. It is therefore hopeless for us to expect to achieve this deeper understanding by reading a few pages of this book—however, we can seek to understand in a measure some of the results which come from a deep study of Analysis.

The number line

We have found it very helpful in earlier sections to represent 'numbers' by points on a line. This helped our intuition, and in fact the ideas of the calculus can also be grasped pictorially by the same representation. Because the picture of the 'number line' is a very accurate one, it is possible to develop the calculus as a useful tool without fully grasping the abstract definition of the set R. Let us once more use the number line as a useful picture on which we can pin our ideas. We previously discussed the number line as a representation for the set Q of rationals. We should now picture it as having all the gaps filled in—with a point corresponding to numbers such as $\sqrt{2}$ and π (see Fig. 4.10). Order on the number line corre-

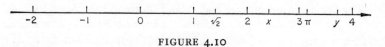

FIGURE 4.10

sponds precisely to the order in R. That is, if $x, y \in R$ and $y > x$, then the point on the line corresponding to the number y will be to the right of the point corresponding to x. Let us now try to understand some further properties of R in terms of the number line.

Every cut gives a number. If you think of an infinite line drawn on paper, you could use a pin to mark on it first the integers, then the rationals with denominator 2, then the rationals with denominator 3, and so on until you had marked on the line all the rational numbers. We would by then have marked an infinite number of points and might appear to have filled up the line—but we have seen that there are other points. If we made the construction of Fig. 4.5 we could take a pair of scissors and 'cut' the line at the point P. Let us consider the effect of cutting our number line at some such point. This will divide R into two pieces—a left piece U and a right piece V—with the property that every real number is in precisely one of the pieces U, V and all the numbers in U are to the left of all the numbers in V. It is very important to be able to say that the 'cut' corresponds to a precise real number c.

It is this property of R which allows in principle for an exact measurement of any physical quantity, like the length of the circumference of a circle, or the weight of a lump of lead.

A bounded set has a smallest bound. Suppose E is a subset of R for which there is some real number b such that:

$$x \in E \Rightarrow x \leqslant b$$

This means that on the number line all of the set E lies to the left of b. Now if E is not empty, it is clear that not every real number b will serve as such an upper bound. Think of sliding b along the line 'as far as you can' to the left while still keeping to the right of E. Obviously if E contains a largest element, *i.e.*, there is $x_0 \in E$ such that $x \in E \Rightarrow x \leqslant x_0$, then this element x_0 is the smallest possible upper bound for E. But not all subsets E have a largest element; for example, suppose E consists of the numbers:

$$x_n = 2 - \frac{1}{n} \quad (n = 1, 2, \ldots)$$

then there cannot be a largest x_n, since each x_n is smaller than the succeeding one x_{n+1}.

Then R has the property that every subset E which is not empty and has some upper bound b has a smallest upper bound s with the properties:

(i) $x \in E \Rightarrow x \leqslant s$.
(ii) $y < s \Rightarrow$ there is some $x \in E$ with $x > y$.

Now (i) says that s is an upper bound, and (ii) tells us that no number less than s is an upper bound.

The set R is complete. In Section 4.4 we spoke of the 'nested interval' property of R as being completeness. Actually, the two properties we have just discussed are really of the same character—they are saying that the points which we feel ought to be there all really exist! All these properties can be formalized in yet another way. Suppose:

$$x_1, x_2, \ldots, x_n, \ldots$$

is any sequence of real numbers which ultimately get arbitrarily close together, then there exists a real number α such that the sequence gets as close as we like to α when n gets large. On the number line,

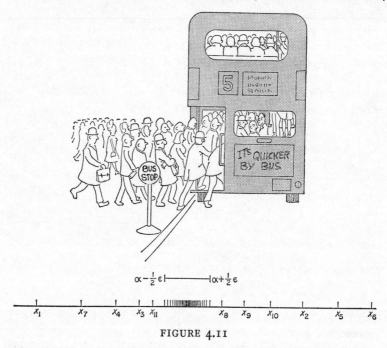

FIGURE 4.11

what we are saying is that a sequence which gets crowded (in the sense that for any length ϵ, however small, we can find an interval of length ϵ, say $\{x \,|\, u \leqslant x \leqslant u + \epsilon\}$, which contains all but a finite number of points in the sequence), must in fact crowd around a unique point α (so that we can take the segment $\alpha - \tfrac{1}{2}\epsilon \leqslant x \leqslant \alpha + \tfrac{1}{2}\epsilon$ always centred at the same point), as shown in Fig. 4.11.

SOME PROPERTIES OF R

A curve and a line cannot cross without intersecting. We have to say what we mean by a curve. Suppose a, b are real numbers with $a < b$, and I is the segment of real numbers $\{x \in R : a \leqslant x \leqslant b\}$ between a and b. We consider the graph of a function $f : I \to R$ which is continuous (that is, the value $f(x)$ is close to the value $f(x_0)$ when x is close to x_0). If we consider the graph of the function—that is, the set

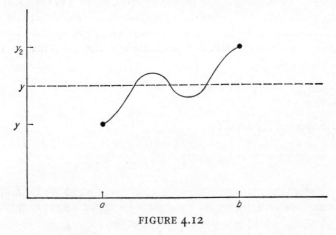

FIGURE 4.12

of pairs $(x, f(x))$ for $x \in I$—we get a continuous curve in the plane. Suppose $y_1 = f(a) < y_2 = f(b)$ and y is a real number with $y_1 < y < y_2$ (see Fig. 4.12). In order to get from $(a, f(a))$ to $(b, f(b))$ the curve has to get from one side to the other of the infinite horizontal line at the height y. Our real numbers are such that this is only possible if there is at least one x with $a < x < b$ and $y = f(x)$. This is another result of the completeness of R—there are no missing points! The truth of this result has far-reaching implications and applications. See Chapter 3 of Burkill, 'A First Course in Mathematical Analysis', for an account of the properties of continuous functions.

Compactness. This is the condition which says that it is not possible to put an infinite set of real numbers in a finite interval $a \leqslant x \leqslant b$ without having an infinite number of them close to some real number in the interval. To put it more precisely, if E is infinite, and $E \subset I = \{x \in R \,|\, a \leqslant x \geqslant b\}$, then we can find at least one real number— call it x_1—such that any interval centred at x_1, however small, will contain an infinite subset of E.

Example 4. Any polynomial function of degree 3 (*i.e.*, a function involving powers of x in which the highest power is x^3) has at least one zero. Suppose:
$$y = f(x) = ax^3 + bx^2 + cx + d$$
defines a function $f: R \to R$, where a, b, c, d are fixed real numbers. Suppose $a > 0$, then if x is large and positive, say $x = k$, then $f(k) > 0$ since the first term will be much larger than the others. Also $f(-k) < 0$. The function f is continuous on the segment $[-k, k]$ and changes from $f(-k) < 0$ to $f(k) > 0$. It must therefore take the value $y = 0$ at least once.

Exercise 4c

1. Suppose, for $n \in N$:
$$x_n = \frac{1}{2} + \frac{1}{4} + \frac{1}{8} + \ldots + \frac{1}{2^n}.$$
Show that the set of all such x_n is bounded above by finding another formula for x_n. Find the least upper bound of this set, and show that it is not a member of the set.
2. George sets out from London at 10 am on Monday and cycles to Cambridge, arriving at 2 pm. The next day he cycles back along the same road leaving at 10 am and arriving at 2 pm. On both journeys he is delayed by traffic and by hills and corners, although the obstructions are not the same in the two directions. Show that there is some point on the road which is reached by George at precisely the same time of day on both journeys. (Assume there are no one-way streets!)

5 | THE REGULARITY OF RANDOMNESS

5.1 What is probability?

Most of us believe statements like:
(i) When you toss a coin there is a 1 in 2 chance that the result will be 'heads'.
(ii) If you throw a die the probability of getting a 'five' is 1 in 6.
What do we understand by such statements and why do we believe them?

 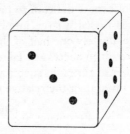

FIGURE 5.1

Perhaps the simplest basis for our belief is the notion that, to be fair, the possible results have to be equally likely. When a coin is tossed we know it never lands on its end, so there are two possible results—heads and tails. The six faces of the die are each marked with spots to distinguish them, so the chance that the face with 5 on it will finish uppermost has to be $\frac{1}{6}$ (see Fig. 5.1). It might be reasonable to try to justify this notion of 'equally likely results' by an appeal to symmetry; but note that the symmetry is only approximate, for there would be no way of distinguishing heads from tails in a perfectly symmetric coin!

A moment's thought will convince the reader that we often use this kind of statement in situations where there is no hope of dividing the outcome into a finite number of equally probable events. For example:

(*iii*) The probability that the Conservatives will win the next general election is 65%.

(*iv*) John Smith has a fifty-fifty chance of obtaining grade B when he sits his examination in A level English next Summer.

In such situations, we are making value judgements using all the relevant knowledge available to us. We are assigning a 'degree of belief' to a statement. In everyday life we often have to make decisions based on such assumptions. For example, John Smith may have to accept or reject conditional offers from universities based on his expectation of a certain result at an A Level examination not yet taken.

Statements (*i*) and (*ii*) are capable of experimental verification; statements (*iii*) and (*iv*) are not. The difficulty is that we cannot repeat a general election under identical conditions a large number of times, nor is it possible for John Smith to keep repeating his A Level English examination. We can keep tossing a coin repeatedly and we do not expect the coin to have any memory. This means that if the coin is tossed a large number of times the result will be heads in approximately one half of the throws. If a given random experiment can result in success or failure, and it is possible to repeat it a large number of times, in such a way that earlier results have no effect on later ones, then the relative *frequency* of success is the ratio:

$$\text{Frequency} = \frac{\text{Number of successes in } n \text{ trials}}{\text{Number of trials, } n}.$$

If the probability of success is p, where p is a real number, $0 < p < 1$, then we expect the frequency of success to be close to p when the number of trials, n, is large.

If we keep repeating the experiment, the frequency, f, will change as n grows. In fact, for each positive integer n, the frequency after n trials can be found, and we can denote it by $f(n)$. This makes $f : N \to [0, 1]$ a function on the integers taking values between 0 and 1.

Now if the experiment is such that p is not known we can estimate it by repeating it a large number of times under identical conditions and observing the frequency:

$$f(n) = \frac{r(n)}{n}$$

where $r(n)$ is the number of times, out of the first n, in which the result is success. As n grows larger we would expect $f(n)$ to get closer and closer to p, so that p is a kind of 'limiting value' for the sequence $f(n)$. As it is impossible to repeat an experiment an infinite number of times in a finite period of time our observations would never fix exactly the real number p.

Sometimes we may think we know what p is, but a repetition of an experiment gives values of $f(n)$ which are not close to p. For example, it would be reasonable to think that:

(v) When a child is born in the United Kingdom, there is a probability of $\frac{1}{2}$ that it will be male.

There is an easy way of observing this experiment a large, but finite, number of times. Write or go to Somerset House, and examine the 'Register of Births' for the last year for which figures are available. Find out the total number of births and the number of male births and calculate the ratio:

$$f(n) = \frac{\text{Number of male births in the year}}{\text{Total number of births in the year}}$$

If this were carried out you might find $f(n) = 0.521$ for a large value of $n > 10^6$. Although this is quite near to $\frac{1}{2} = 0.5$, it differs from $\frac{1}{2}$ by more than you would expect with such a large sample. What should our reaction be? We do not discard the whole notion of frequency, but instead conclude that the figure $\frac{1}{2}$ in statement (v) is very likely to be inexact. The probability of a male birth is some fixed real number p, $0 < p < 1$. We do not know what p is, but experimental evidence indicates it is about 0·52.

Exercise 5a

As an experiment, try finding $f(n)$ from the birth columns of each of three papers (one local and two national) over a period of, say, 3 weeks. Do they differ from each other and from 0·52? Think out possible reasons for the differences you observe.

We shall give details in Sections 5.4 and 5.5 of an actual experiment, which can be performed without difficulty, to give some experience of the behaviour of the frequency $f(n)$ after n trials.

5.2 Combining probabilities

In order to make our ideas precise, let us consider an experiment whose result is an element x of some space S (see Section 1.3). The choice of the space S has to be appropriate for the experiment. For example:
   (*i*) For coin-tossing, S contains two points H, T.
   (*ii*) For throwing a simple die, $S = \{1, 2, 3, 4, 5, 6\}$.
   (*iii*) For measuring the weight of a newborn baby in some unit we could have $S = N_+ = \{0, 1, 2, \ldots\}$ if we always measure to the nearest whole number of units, or we could have $S = R$ if we wanted to measure exactly.
   (*iv*) For recording the result of an A Level examination, the space $S = \{a, b, c, d, e, f, g\}$.

Having decided on S, we may not be interested in the precise outcome x of the experiment, but rather in some property of x. For example if we toss a die we might be interested in whether the result x was even. The set of possible even results is:

$$E = \{2, 4, 6\} \subset S = \{1, 2, 3, 4, 5, 6\}.$$

In general, if E is a subset of S, we are interested in whether or not $x \in E$. In a random experiment we would like to assign a probability to all subsets E in S. This is possible if S is finite or can be arranged as a sequence $\{x_1, x_2, \ldots, x_n, \ldots\}$, but not otherwise. For example, it is not possible if S is the set R of real numbers (see Chapter 8).

The subsets $E \subset S$ to which we can assign a probability, $P(E)$, are called *events*. If we carry out the experiment and obtain a result x belonging to E, we say that the event E happens; while if x belongs to E', the complement of E, we say that E does not happen. For example, if we measure the weight of a newborn baby in kilogrammes, the event $\{x > 4\} = E \subset S$ would be of interest. This event happens if and only if the baby weighs more than 4 kilogrammes.

If we are interested in events E, F in a space S, then we are automatically interested in:

E', the complement of E
$E \cup F$, the event that at least one of E, F happen
$E \cap F$, the event that both E, F happen
$\emptyset$, the event that nothing happens
S, the event that something happens

COMBINING PROBABILITIES

[These operations on sets were discussed in Chapter 1. The Venn diagram, Fig. 5.2, is a helpful way of representing events.]

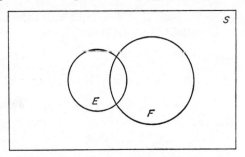

FIGURE 5.2

When S is finite or a sequence we assume that all subsets are events. For 'larger' spaces we must assume that the operations of taking complement, union and intersection when applied to events always yield events. For simplicity, let us restrict attention to the case of a finite space S. Then every subset $E \subset S$ will be an event. If $P(E)$ denotes the probability that E happens, then:

Law 1 $P(\emptyset) = 0, P(S) = 1$

This simply says formally that when the experiment is carried out something certainly happens. If our intuitive picture means anything, then the probabilities of related events must be related. The first relation is obvious.

Law 2   If $E \subset F$, then $P(E) \leqslant P(F)$

For the event E cannot happen without F happening, so F must be at least as likely as E.

If A, B are subsets of S with no common point we say they are disjoint. As events we call them *mutually exclusive*: that is, if A happens then B cannot happen. In this case we have:

Law 3   If $A \cap B = \emptyset$, then $P(A \cup B) = P(A) + P(B)$

Although Law 3 seems intuitively reasonable, let us check it by using the frequency interpretation. Repeat the experiment a large number, n, of times and count:

$r(A)$ = the number of times A happens

$r(B)$ = the number of times B happens
$r(A \cup B)$ = the number of times A or B happen

As it is impossible for both A and B to happen together:
$$r(A \cup B) = r(A) + r(B)$$
Dividing by n gives:

Frequency of $A \cup B$ = frequency of A + frequency of B

But the frequency of an event is approximately equal to its probability when n gets large. It follows that the probabilities must satisfy Law 3.

An immediate application of Law 3 is:
$$P(E) + P(E') = 1,$$
i.e., for any event E, the probability that E happens must be 1 minus the probability that E does not happen.

In the case of a finite space S (see Fig. 5.3), the easiest way to assign $P(E)$ for all events is to define it for the one-point events. Suppose:
$$S = \{x_1, x_2, x_3, \ldots, x_n\}$$
and
$$P\{x_r\} = p_r \quad (r = 1, 2, \ldots, n).$$

Then since the events $\{x_1\}, \{x_2\}, \ldots, \{x_n\}$ are disjoint and their union is S, Law 3 gives $p_1 + p_2 + \ldots + p_n = 1$, and we can find the probability of any event E by simply adding the probabilities of each of the *points* in E. In Fig. 5.3, E contains x_2 and x_5, so $P(E) = p_2 + p_5$.

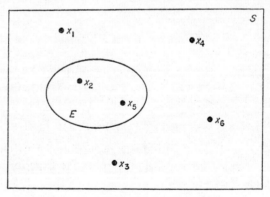

FIGURE 5.3

COMBINING PROBABILITIES 83

Example 1. Suppose the experiment consists of throwing two dice, a red one and a green one. The result is then (a, b) where a is the figure on the red die, and b the figure on the green die. If the dice are symmetrical, it would be reasonable to assign equal probabilities to each of the 36 points in S'.

(1, 1), (1, 2), (1, 3), (1, 4), (1, 5), (1, 6), (2, 1), (2, 2), (2, 3),
(2, 4), (2, 5), (2, 6), (3, 1), (3, 2), (3, 3), (3, 4), (3, 5), (3, 6),
(4, 1), (4, 2), (4, 3), (4, 4), (4, 5), (4, 6), (5, 1), (5, 2), (5, 3),
(5, 4), (5, 5), (5, 6), (6, 1), (6, 2), (6, 3), (6, 4), (6, 5), (6, 6).

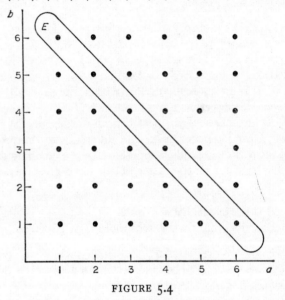

FIGURE 5.4

In Fig. 5.4 we represent S as a co-ordinate diagram, which shows that for each value of b, there are 6 values for a. The event E on the diagram is the event that 'the sum of figures on the dice is 7'. Clearly E contains 6 of the possible points, each with probability $\frac{1}{36}$, so $P(E) = \frac{6}{36} = \frac{1}{6}$.

Exercise 5b

1. In the experiment of Example 1, calculate the probability of the events:
    (a) The sum of the figures on the dice is 3.

(b) The sum of the figures on the dice is 12.
(c) The sum of the figures on the dice is 13.
(d) The two dice show the same figure.
(e) The red die shows a figure 2 more than the green die.

2. Show that, in any probability space, for events A, B we always have:

$$P(A \cup B) + P(A \cap B) = P(A) + P(B).$$

5.3 Independent events

The richness of the theory of probability results from introducing an additional concept to the structure discussed in Section 5.2. Given two events A, B we can ask whether the knowledge that A has occurred makes it more likely or less likely that B will occur. For example, if A, B are mutually exclusive, then the probability that B occurs, given we know that A has happened, is zero. We say that the events A, B are *independent* (or statistically independent) if knowledge that A has happened makes no difference to the probability that B happens. For example, suppose we were examining the earnings of people living in Doncaster. Then for a given person x we could say that:

A is the event that x is male
B is the event that x earns more than £3000 pa.

Taking account of the structure of British society we should not expect these events A, B to be independent, for it is more likely for a man to have a larger salary than a woman. However, for the same inhabitants of Doncaster we could consider:

C is the event that x has blue eyes.

Now it seems reasonable to expect that males and females are equally likely to have blue eyes, so that the knowledge that someone is a male does not affect the probability of having blue eyes.

To see what this means we could look at the whole population of Doncaster and record for each person whether or not they are male, whether or not they earn more than £3000 pa, and whether or not they have blue eyes. We could analyse the results in what are called *contingency tables* (see Tables 5.1 and 5.2).

INDEPENDENT EVENTS

TABLE 5.1 Inhabitants of Doncaster by income

	Income, £pa		
	>3000	<3000	Total
Male	1205	52 025	53 230
Female	365	53 950	54 315
Total	1570	105 975	107 545

TABLE 5.2 Inhabitants of Doncaster by colour of eyes

	Colour of eyes		
	Blue	Not blue	Total
Male	40 105	13 125	53 230
Female	40 811	13 504	54 315
Total	80 916	26 629	107 545

We should think of our random experiment as 'picking a person in Doncaster at random' so that all 107 545 inhabitants are equally likely. Then:

$$P(A) = \tfrac{53\,230}{107\,545} \approx 0.495$$
$$P(B) = \tfrac{1570}{107\,545} \approx 0.015$$
$$P(C) = \tfrac{80\,916}{107\,545} \approx 0.752$$
$$P(A \cap B) = \tfrac{1205}{107\,545} \approx 0.011$$
$$P(A \cap C) = \tfrac{40\,105}{107\,545} \approx 0.373$$

for we can calculate the probability of an event E by taking the proportion of points for which E happens. Now it is easy to check that:

$$P(A).P(B) \neq P(A \cap B)$$
and
$$P(A).P(C) = P(A \cap C).$$

We can understand this formally in terms of the frequency interpretation.

Given a large number of n trials, put:

$r(A)$ = number of times A occurs
$r(B)$ = number of times B occurs
$r(A \cap B)$ = number of times both A and B happen

Then if we only count those trials at which A happens:
$$\frac{r(A \cap B)}{r(A)}$$
is the proportion of these at which B also happens. For large n we would expect $r(A \cap B)/r(A)$ to be close to the probability that B happens, given that we know A has happened. Now:
$$\frac{r(A \cap B)}{r(A)} = \frac{r(A \cap B)}{n} \Big/ \frac{r(A)}{n}$$
which is approximately:
$$\frac{P(A \cap B)}{P(A)}$$
Hence A, B are independent events if, and only if:
$$P(B) = \frac{P(A \cap B)}{P(A)}$$
that is:
$$P(A).P(B) = P(A \cap B)$$

Note that if A, B are disjoint, then $P(A \cap B) = 0$, so that A, B will not be independent unless at least one of them has zero probability. We should think of the relation of independence as a kind of mixing condition on two events A, B.

Example 2. Figure 5.4 illustrates the result of throwing two dice. In this S, suppose:
$$A = \{\text{red die shows even number}\}$$
$$B = \{\text{green die shows 6}\}.$$
Then $(a, b) \in A$ if and only if $a = 2, 4$ or 6 and b has any value. Hence:
$$P(A) = \frac{3.6}{36} = \frac{1}{2}.$$
$(a, b) \in B$ if and only if $b = 6$ and a has any value. Hence:
$$P(B) = \tfrac{6}{36} = \tfrac{1}{6}.$$
$(a, b) \in A \cap B$ if and only if $a = 2, 4$ or 6 and $b = 6$. Hence:
$$P(A \cap B) = \tfrac{3}{36} = \tfrac{1}{12}$$
$$= P(A).P(B)$$
so that A, B are independent.

In this example the above situation is typical; if A is any event in which only the red die is restricted, and B is any event in which only the green die is restricted, then A, B are independent.

Now consider the events:

E = {sum of figures on the two dice is 7}
C = {the red die shows a figure 3 more than the green die}

Then $P(E) = \frac{1}{6}$, as we saw in Example 1. C contains the three points (4, 1), (5, 2), and (6, 3), and $E \cap C$ contains just one point (5, 2). Hence $P(E \cap C) = \frac{1}{36}$, $P(C) = \frac{3}{36} = \frac{1}{12}$ and $P(E).P(C) \neq P(E \cap C)$. The events E, C are not independent even though they are not disjoint.

This notion of independence is important in making precise the idea we have used more than once that we can keep repeating an experiment in such a way that earlier results have no effect on the later ones—the experiment has no built-in memory.

Example 3. Suppose we are tossing a fair coin, and we want to know the probability of two heads and a tail, in that order. So the appropriate space is $S = \{H, T\}$, and $P\{H\} = \frac{1}{2} = P\{T\}$. Then if we toss it three times independently, the appropriate space is $S \times S \times S$ which has eight points in it:

TTT, TTH, THT, HTT, THH, HTH, HHT, HHH.

Since we are assuming independence of the throws:

$$P\{HHT\} = P\{H\}.P\{H\}.P\{T\} = \tfrac{1}{8}.$$

In fact each of the points in $S \times S \times S$ will have the same probability $\frac{1}{8}$.

This notion of independence implies some results which may run counter to popular belief. Suppose you toss a fair coin 10 times and get heads each time, is it better to bet on heads or tails for the next throw? Most people who answered without much thought would say 'tails', since by 'the law of averages', heads and tails should each appear about the same number of times. Actually the independence assumption implies that the probability of heads at the eleventh throw is not affected by the knowledge of 10 preceding 'heads'—so it is still $\frac{1}{2}$.

In fact if you tossed a coin 10 times and obtained 10 heads, you would begin to suspect that the coin was biased—and therefore the correct strategy would be to bet on another head at the eleventh throw. Really what we are saying here is that if, in an actual experiment the frequency $f(n)$ of heads approaches 0·65 as n gets large, we do not decide that probability theory is nonsense, but rather that the coin is not fair—it must be biased so that heads are more likely than tails. We need to be careful in this kind of deduction for in any fixed number n of repetitions (no matter how large) there is a small chance of observing a value of $f(n)$ which differs from 0·5 by as much as 0·15 even if the coin is fair. We can never be certain the coin is biased as a result of a finite experiment. One of the objects of statistics is to say how confident one can be about conclusions of this type.

Example 4. Consider now any probability space S with an event $E \subset S$ such that:
$$P(E) = p, \quad P(E') = 1 - p = q.$$
Suppose we repeat the random experiment again and again, recording each time whether E happens or does not happen. Assume independence between the repetitions. What is the probability that, in a sequence of n repetitions, the event E will occur r times?

To solve this, record the result of n trials by a succession of n digits 0 or 1:
$$1100010101\ldots 11$$
where we write 1 when we observe that the event E has happened and 0 when E does not happen. Since there are 2 choices for each digit, there must be 2^n possible choices of n digits. By independence, the probability of one such sequence is
$$ppqqqpqpqp\ldots pp$$
where we write $p = P\{E\}$ each time 1 appears and $q = P\{E'\}$ each time 0 appears. Hence if p appears r times, q will appear $(n-r)$ times and the probability of that particular result will be $p^r q^{n-r}$, irrespective of the order of appearance of the factors. The number of possible sequences of n digits 0, 1 in which there are precisely r of the digits 1 is the same as the number of ways of picking r objects out of a collection of n objects (since we can pick the places for 1 and fill up the rest with 0). We saw in Chapter 3 that this was:
$$\binom{n}{r} = \frac{n!}{r!\,(n-r)!} = \frac{n(n-1)\ldots(n-r+1)}{r(r-1)(r-2)\ldots 2.1}$$

Hence the probability of observing the event E, r times out of n independent trials is:

$$\binom{n}{r} p^r q^{n-r}.$$

Note that we have to observe it some number k times, where k is some integer between 0 and n. This means that if we add the corresponding probabilities we get

$$p^n + np^{n-1}q + \ldots + \binom{n}{r} p^r q^{n-r} + \ldots + q^n = (p+q)^n = 1^n = 1$$

by the binomial theorem (see page 48, Example 5 in Chapter 3).

Exercise 5c

1. How many times do you need to toss a fair coin to ensure that the probability of getting at least one head is:
   (a) greater than $\frac{99}{100}$ (b) greater than $\frac{999}{1000}$
2. Suppose that, of the students who cheat at examinations, 1 out of 10 is caught. What is the probability that a person who cheats 10 times gets caught at least once? What if he cheats 20 times? (Hint: find the probability of cheating and not being found out.)

5.4 A group experiment

We have discussed briefly part of the theoretical basis for probability arguments. Before going any further, the reader is urged to spend a couple of hours actually repeating, many times, a simple random experiment and recording and analysing the results. More will be learnt if each individual in a group carries out the experiment and the results are finally collated and compared.

The only apparatus required is a collection of convex shapes made from stiff cardboard or plastic. (There are several toys designed for toddlers which contain suitable pieces ready made.) There should be at least four identical pieces of each kind, and each member of the group should use just one piece—the number of different shapes required depends on the size of the group. It is desirable to use shapes which differ both in size and design—Fig. 5.5 illustrates a typical collection obtained from a child's toy.

We also require a set of parallel lines spaced apart at an equal distance, which must be larger than the diameter† of the pieces being used so that it is impossible for any one piece to intersect more than one of the lines. If you have a floor with hardwood 'strip' flooring, this forms a ready-made grid of equally spaced parallel lines which is

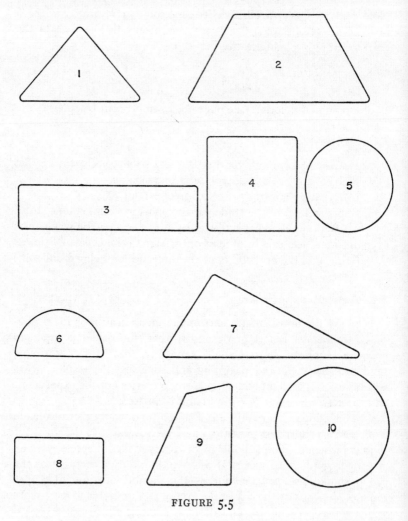

FIGURE 5.5

† For any figure the diameter is the longest distance separating two points of the figure. This generalizes the notion of diameter for a circle.

A GROUP EXPERIMENT

suitable provided you choose sufficiently small shapes. If there is no suitable floor, each experimenter will need to prepare a large sheet of paper on which the parallel lines are drawn.

Each experimenter should take the piece allotted to him and first describe it as fully as possible, making measurements and recording at least the information in Table 5.3, which resulted from using shape number 3.

The piece should be dropped carelessly (that is, without looking at the floor) from a height onto the grid of parallel lines on the floor (Fig. 5.6). It will bounce, slide, roll, and eventually come to rest on the grid. We say that:

(i) the event E happens if the piece cuts one of the lines
(ii) the event E does not happen if it lies completely between two lines

FIGURE 5.6

Keep repeating the trial throw at least 512 (= 2^9) times, but preferably 1024 (= 2^{10}) times and record the results by writing down the digit 1 each time E happens, and the digit 0 each time E does not happen. This experiment is very simple, but it is only worth doing if the results are recorded accurately.

TABLE 5.3 Typical result

SHAPE USED No. 3. Rectangle 5 cm × 1·2 cm.
AREA 6·0 sq cm.
PERIMETER 12·4 cm.
DIAMETER 5·2 cm.
DISTANCE BETWEEN LINES 6·15 cm.

Result of experiment

Number of trials	2	4	8	16	32	64	128	256	512	1024	2048
Number of successes	1	3	4	10	24	45	77	164	333	680	1351
Frequency	0·50	0·75	0·50	0·62	0·75	0·70	0·68	0·64	0·65	0·66	0·66

Our intuitive feeling is that there ought to be a number p, $0 < p < 1$, such that $P(E) = p$. If for each integer n we count the number of digits 1 among the first n digits in our sequence and find there are r '1's this means that the proportion of times (among the first n) that our experiment resulted in the event E, which we think of as success, is $f(n) = r/n$.

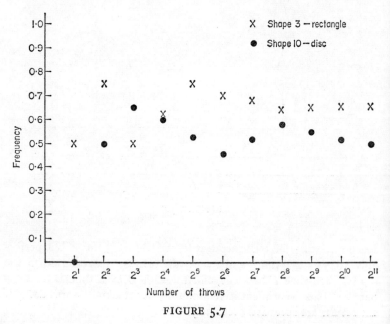

FIGURE 5.7

WHAT DETERMINES p?

We would like to plot a graph of this frequency $f(n)$ against n to see how it behaves as n varies. The scale is difficult since the range for n is so large, and a clearer picture results if we calculate $f(n)$ for the special integers $n = 2^q$, and plot a graph of $f(2^q)$ against q. If four people have been using identical pieces and each has done the experiment 2^{10} times, it is possible to get two more points on the graph by first treating the results of two people as a single experiment with 2^{11} trials, and then the combined results from four people as an experiment with 2^{12} trials. Figure 5.7 illustrates the graphs of two such random experiments. Each individual in the group should prepare his own table (as in Table 5.3) and plot his own graph, obtaining two extra points by using the results of 3 others. It will be noticed from the graphs that the frequency changes erratically in the earlier stages of the experiment, but that before too long it settles down and varies very little when the number of trials is large. The last point on the graph gives the best estimate of the hypothetical value of p we have postulated.

5.5 What determines p?

In the above experiment each shape has its own value p for the probability of cutting a line when it is dropped at random. We do not know what p is but the experiment with 2^{11} trials gives us quite a good estimate of its value. This estimate of p should be good enough

TABLE 5.4 Comparison of results
(Distance between lines = 6·15cm)

Shape number	Area in sq cm	Perimeter in cm	Diameter in cm	Final frequency
1	3·1	8·4	3·4	0·45
2	8·7	12·6	5·0	0·66
3	6·0	12·4	5·2	0·66
4	6·3	10	3·5	0·54
5	5·0	7·8	2·5	0·42
6	2·5	6·4	2·5	0·32
7	6·6	13·2	5·6	0·65
8	3·0	7·4	2·8	0·40
9	4·7	9·1	3·0	0·49
10	7·5	9·7	3·1	0·50

to allow us to guess which properties of the shape are relevant to the probability. It helps to prepare a table showing the results obtained by a class for all the shapes used (see Table 5.4, which was obtained using the shapes 1–10 of Fig. 5.5).

Look in your table for shapes in which two of the measurements are nearly the same and the third quite different, and note the effect on the observed frequency. You ought to be able to decide that only one of the three measurements tabulated is relevant to the probability p. Which one is it?

There is one shape for which it is almost obvious what p should be —that is the circle. Suppose we have a circle of diameter d thrown on a grid of lines at a distance h apart, where $h > d$; then the circle will cut a line if, and only if, the distance from its centre to the nearest line is less than the radius, $\frac{1}{2}d$. The possible values of this distance vary from 0 to $\frac{1}{2}h$ and all are 'equally likely'. Hence we obtain p as the ratio:

$$p = \frac{\frac{1}{2}d}{\frac{1}{2}h} = \frac{d}{h}$$

of the size of the interval which results in the event E to the total size possible.

If one of the shapes thrown is a circle, (numbers 5 and 10 in Fig. 5.6), then we can compare the observed frequency after a large number of trials with this theoretical answer $p = d/h$, and we should find the two numbers are approximately the same. If we have succeeded in discovering which property of the shape determines p, we can calculate a theoretical probability for each shape and compare this with the final frequency observed. It is possible to give a mathematical argument leading to the value of p for any convex shape, and this is summarized in an Appendix at the end of this Chapter.

5.6 Law of large numbers

Our experience in carrying out a random experiment should have confirmed our intuition that the abstract idea:

(i) 'An event E happens with probability p' has got a valid 'frequency' interpretation; namely

(ii) If we repeat the experiment a large number, n, times and count the number of times, r, that E happens, then the ratio r/n will be approximately p.

LAW OF LARGE NUMBERS

We saw in Example 3 that, by assuming independence between repetitions of the experiment, we could calculate precisely the probability of observing r successes in a sequence of n trials. This was:

$$\binom{n}{r} p^r q^{n-r}.$$

If we ask what is the probability of obtaining an observed frequency after $n = 1024$ trials which differs from the theoretical value of p by less than 0·01, then we have to add together terms of this type for values of the integer r such that:

$$p - \frac{1}{100} < \frac{r}{n} < p + \frac{1}{100}.$$

For $n = 1024$ this means calculating 21 terms like:

$$\binom{1024}{360}(0\cdot 35)^{360}(0\cdot 65)^{664}$$

and adding them together. This would be a formidable task even with the aid of a calculating machine! Theoretical methods of obtaining a very accurate answer to this sum have been developed (and would give a value of 0·75 in this case). With 2^{10} trials the relative frequency is still a 'random' number and our calculation shows that it will be accurate to within $\frac{1}{100}$ with probability about 0·75. If we want greater certainty of such an accuracy we would have to increase the number of trials—for example with 2^{13} trials the frequency will be within $\frac{1}{100}$ of p with a probability of about $\frac{199}{200}$. This illustrates what is called the *weak law of large numbers*, which makes precise the statement (*ii*) above. It can be stated as follows:

Given any prescribed error $\epsilon > 0$, and desired certainty c, $0 < c < 1$, we can find an integer N such that if we carry out n trials ($n \geqslant N$) and count the number r of times that E occurs, the frequency r/n will be within ϵ of p with a probability of at least c.

This is not the end of the story. The frequency interpretation can be given even greater precision by constructing a mathematical model for repeating the experiment not only a large number of times, n, but infinitely often. Suppose we have such a model (it would take us too long to describe its construction, though it extends the notions we have developed for finite n) and suppose we calculate the observed frequency $f(n)$ for each integer n. This means we have a *random sequence*:

$$f(1), f(2), \ldots, f(n), \ldots$$

The weak law of large numbers says that when n is large $f(n)$ is likely to be close to p. This is a statement about each particular large integer n. We would like to make precise the idea that $f(n)$ actually gets closer to its theoretical value p as n gets larger and larger. Our experience tells us that $f(n)$ does not get closer to p steadily, even when n is quite large, so we have to be careful about the precise formulation. The result is called the *strong law of large numbers*, and we can express it as follows:

> Given any prescribed error $\epsilon > 0$, there is probability 1 that there will be an integer N (depending on the sequence) such that the difference $(f(n) - p)$ will have an absolute value:
>
> $$|f(n) - p| < \epsilon,$$
>
> for all of the integers $n \geqslant N$.

This means that $f(n)$ is not only close to p, but stays close to p as n continues to grow, and this event surely happens. This is the sense in which random events behave regularly. Although the event is random and even the calculation of the frequency $f(n)$ of success for a particular large n is also a random number, which may not even be close to p, yet ultimately $f(n)$ stays close to the probability p. There is an amazing regularity even in random events.

5.7 Application

We are led to believe from various experiments that the air around us consists of an enormous number of molecules (10^{24} in every cubic centimetre) each moving incessantly. Since in practice they keep colliding and changing velocities and directions, the motion of any particular molecule can be thought of as random. It is conceivable that this random motion might result in all 10^{28} molecules in the room suddenly moving in the same direction away from us. This would be catastrophic for it would leave us with no air to breathe! We know this doesn't happen—and the law of large numbers provides an explanation. Since we can think of the molecules as behaving independently 'at random' the proportion moving to the right or to the left at the time of observation is always going to be close to $\frac{1}{2}$. The small variations (very small indeed since 10^{28} is rather a large number of trials!) from this proportion of $\frac{1}{2}$ cannot be detected and we do not run out of air!

There is an elaborate theory of statistical mechanics which depends for its validity on this regularity in randomness. This regularity ensures that in many cases one can give mechanistic or statistical models for a given situation, and have both models valid. For if the strong law of large numbers is applicable then the statistical model leads to the prediction that an event will always happen—and this is equivalent to a mechanistic model in which the non-happening of the event is not conceived.

APPENDIX

The probability of cutting a line of a grid

Suppose we have a parallel grid of lines whose distance apart is h, and suppose we drop a straight segment of length l onto the grid and count the number of intersections between the grid and the segment. This gives a random integer N with possible values from 0 to r. For a fixed segment, if we repeat a large number, k, of times we would expect to have:

0 intersections about kp_0 times
1 intersection about kp_1 times
.
.
.
r intersections about kp_r times

so that the total number of intersections would be about:

$$k(0p_0 + 1p_1 + 2p_2 + \ldots + rp_r)$$

where $p_i = P\{N = i\}$, $i = 0, 1, \ldots, r$, and r is the largest integer such that the line segment of length l can have r intersections with the grid. (In fact r is the smallest integer bigger than the ratio l/h.)

Thus the *average* number of intersections would be:

$$E(N) = p_1 + 2p_2 + \ldots + rp_r.$$

This number is called the *expectation* of the random integer N. Now if we double the length l we shall double $E(N)$, for we could observe separately the two halves of the new longer segment. It follows that, for fixed grid distance h, $E(N)$ must be proportional to the length l. This means that there is some real constant d such that:

$$E(N) = dl.$$

Further, if we double h, this has the same effect as dividing l by 2, so $E(N)$ must be inversely proportional to h. It follows that there is an absolute constant c such that:

$$E(N) = c\frac{l}{h}.$$

This argument extends now to any polygonal figure dropped at random on the grid. For each side of length l_i the expectation of the number N_i of intersections is cl_i/h. Hence if l is the perimeter of the polygon, the expectation of the number of intersections by the boundary of the figure is $\frac{c}{h}(l_1 + \ldots + l_k) = cl/h$

Now suppose the figure is a convex polygon of diameter $d < h$. Then when it is thrown the number of intersections is either 0, 1 or 2 (see Fig. 5.8), since at most one line of the grid is involved. Now the

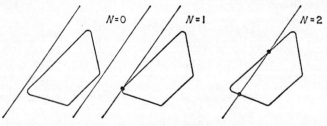

FIGURE 5.8

case $N = 1$ corresponds to the shape landing so that it touches a line exactly but does not cross it. Since all possible positions relative to the line are equally likely, the probability of this event can only be zero. If p is the probability that $N = 2$ then:

APPENDIX

$$E(N) = 2P\{N = 2\} + 1 \cdot P\{N = 1\} + 0 \cdot P\{N = 0\} = 2p.$$

Hence:

$$p = \frac{c}{2}\frac{l}{h}$$

where $\frac{1}{2}c$ is a real constant, not dependent on the figure. Since any convex shape can be approximated by a convex polygon, this result is also valid for any shape.

This means that, if our analysis is valid, we have proved that the probability of success in our experiment is directly proportional to the ratio l/h, where l is the perimeter of the shape and h the distance between lines. But for a circle of diameter d, the perimeter is πd, and:

$$p = \frac{c}{2}\frac{\pi d}{h} = \frac{d}{h}$$

so that

$$\frac{c}{2} = \frac{1}{\pi}.$$

We have thus proved that the theoretical probability of success is:

$$p = \frac{1}{\pi}\frac{l}{h}$$

where l is the perimeter and h is the distance between the parallel lines.

It is worth remarking that there is a classical result of this type, called the 'Buffon needle' experiment. This consisted in dropping a needle on a parallel grid—the needle being thought of as a segment with no thickness. If we take the 'perimeter' of the needle, it will be twice its length λ, because we are thinking of the needle as a convex figure with two sides in the same place, so that:

$$p = \frac{2}{\pi}\frac{\lambda}{h}$$

where h is the distance between the lines. The experiment considered in Sections 5.4 and 5.5 is thus a generalization of the classical needle experiment. Note that it can be used as a method of estimating the value of π. Why is it not a good method of finding π?

6 | THE GAMBLER'S RUIN

6.1 Gambling in a fair game

How often do habitual gamblers find themselves bankrupt? The situation called 'gambler's ruin' is that reached by a gambler who cannot continue to stake any money because his assets are completely exhausted. We will not be interested in the morals of the situation, but rather in the mathematical laws which govern it.

FIGURE 6.1

In real life, if you gamble in a casino or gaming house or play with the so-called 'one-armed bandit', the game is not fair; it is weighted in favour of the casino. We shall look briefly at these unfair games later, but first let us examine the very simplest example of a fair game played between two people whom we shall call Peter and Paul. They play a simple game of chance and agree that each time Peter wins, Paul will pay him £1; each time Paul wins, Peter will pay

GAMBLING IN A FAIR GAME

him £1. At the beginning of the session Peter has a capital of £n and Paul a capital of £m. The contest ends when one or other of the players is ruined. The game is called 'fair' if Peter and Paul are equally likely to win—in the terms of Chapter 5, the probability of the event that Peter wins a play is $\frac{1}{2}$, and the probability of a win for Paul is also $\frac{1}{2}$.

For this situation there is a simple answer to the question 'Which of the players is more likely to be ruined?' To find the answer let us picture the progress of the gambling session as a *random walk* on the integers. We should picture a random walk as a progress on the number line in which you step only on integer points, and at any time you have an equal chance of moving one step to the right or one step to the left. It is sufficient to look at the progress of Paul's fortune, for the total amount of money remains constant at $m + n = a$, say. If Paul's fortune ever reaches 0 then he is ruined, while if it ever reaches a then Peter is ruined. If at the beginning of a play, Paul's fortune is z, then at the end of it, it will be $z + 1$ if he wins and $z - 1$ if Peter wins, each of these events having probability $\frac{1}{2}$. If we mark the integers from 0 to a on the number line, then for a play starting at z, the player's fortune moves to the right one step to $z + 1$ with probability $\frac{1}{2}$ or to the left to $z - 1$ with probability $\frac{1}{2}$ (see Fig. 6.2).

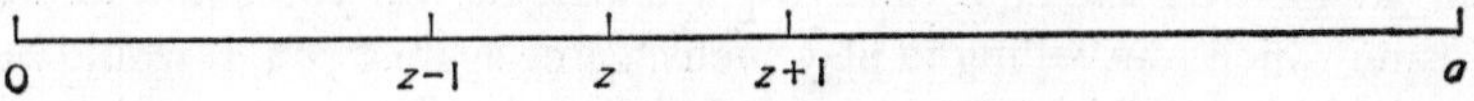

```
L_____|_____|____|____|_____J
0     z-1      z    z+1                        a
```

FIGURE 6.2

It is assumed that successive games are independent, so the probability of moving to the right or the left is always $\frac{1}{2}$ no matter what has happened previously. This motion on the integer points is called a symmetric random walk: it is said to have absorbing barriers at 0 and a because the motion has to cease as soon as one player is ruined. The question of which player is ruined now is the question of which boundary is reached first.

For $0 \leq z \leq a$, let q_z denote the probability of reaching 0 before a, starting from z. Then

$$q_0 = 1, \; q_a = 0 \tag{6.1}$$

for if $z = 0$ the walk is already at 0, whereas if $z = a$, the walk will never leave a. Starting from z we have seen that it moves to $z + 1$ and

$z-1$ each with probability $\frac{1}{2}$; this is its first step to the boundary. Hence:

$$q_z = \tfrac{1}{2}(q_{z+1} + q_{z-1}) \qquad (1 \leqslant z \leqslant a-1)$$

This kind of equation is called a *recurrence relation*; it is solved by:

$$q_z = c + dz$$

where c and d are any real numbers. Substituting $q_z = c + dz$ in equation (6.1) and solving for c and d gives $c = 1$, $d = 1/a$ so that:

$$q_z = 1 - \frac{z}{a} \qquad (0 \leqslant z \leqslant a)$$

as the unique solution of this system of equations.

Hence, if Paul starts with £m and Peter starts with £n:

Paul is ruined with probability $\qquad 1 - \dfrac{m}{m+n} = \dfrac{n}{m+n}$

Peter is ruined with probability $\qquad \dfrac{m}{m+n}$

This shows that even in a fair game, the rich gambler has an advantage over the poor one. In particular, even if you could find a casino which was willing to play such a game against you, it would be inadvisable to continue playing it indefinitely. For you would have only some finite capital to play with, while the casino would have in effect unlimited funds—and this means that if you continue to play against the casino you are almost certain to be ruined eventually.

Actually, a gambler has a device open to him to overcome this problem. He could decide, before starting, to limit his potential winnings. For example, suppose Paul starts with £80 and plays the fair game against a casino which is infinitely rich, he could decide that if his total wealth ever increased to £100 he would steadfastly walk out of the casino with the money in his pocket. That is, Paul can opt to stop playing when, and if, the circumstances are favourable to him. With this strategy he has effectively made it into a game against an opponent with £20 so that there are two possible outcomes:

(1) Paul will lose all his money with probability $\dfrac{20}{20+80} = \dfrac{1}{5}$

(2) Paul will increase his capital from £80 to £100 with probability
$$\frac{80}{20+80} = \frac{4}{5}$$

Before leaving the 'fair game' situation, let us ask a question about 'runs of good luck'. Even those who do not gamble may play a game like bridge for hours on end. It is common experience that on some such occasions you feel that the cards are consistently against you, while on others you feel that on the whole they definitely favour you. Can this be explained in terms of the mathematical theory? It is not easy to analyse the game of bridge because the number of possible outcomes of a single hand is enormous, so let us ask the question about the simple fair game which we have already considered.

Suppose then that Peter and Paul have unlimited capital (so that neither will be ruined), and that they play their game regularly once every second for 12 hours—that is, a total of 43 200 times. They keep a record of the history of the game so that after n trials they know whether Paul has won more often than Peter, so that he has positive nett winnings, or whether Peter has won more often than Paul in the first n trials. We say that Peter is more fortunate than Paul if he is ahead for more than half the time. Then it is possible to calculate the probability of the events:

E_1 : less fortunate player ahead for less than 2 hours
$P(E_1) = 0.52$
E_2 : less fortunate player ahead for less than 1 hour
$P(E_2) = 0.38$
E_3 : less fortunate player ahead for less than 10 min
$P(E_3) = 0.15$
E_4 : less fortunate player ahead for less than 1 min
$P(E_4) = 0.04$

The method of calculating these probabilities is called 'the first arc sine law' because it has the form:

$$P\{Z < \alpha\} = \frac{4}{\pi} \sin^{-1} \alpha^{\frac{1}{2}} \qquad (0 < \alpha < \tfrac{1}{2})$$

where Z is the proportion of time the less fortunate player is in the lead. The reader who wants to understand the justification of this formula (which requires a fair amount of technique) is referred to Chapter 3 of Feller's, 'Introduction to Probability and its Applications'.

These results are somewhat surprising. For they show that there is a probability of about $\frac{1}{25}$ that the less fortunate of the players will be ahead for only 1 min in 12 hours' play, so that we should expect this to happen about once if the operation is repeated 25 times. Thus, our runs of so-called good luck or bad luck are in fact completely predicted by the mathematical laws of probability.

Exercise 6a

The following table indicates various possible starting capitals in a two-person fair game. Complete it by calculating the probabilities of ruin for each of the players.

Paul's starting capital	Peter's starting capital	Probability of ruin for Paul	Peter
9	1		
90	10		
950	50		
5	95		
2000	8000		

6.2 Strategy in an unfair game

In real life, casinos differ from the model we have described in two ways. Firstly the game is not normally so simple—there are usually several possible outcomes with differing amounts of gain; and secondly, it is not usually fair—that is, the probabilities are arranged to be favourable to the house, so that what is called the expectation of gain by a player is negative. Provided the possible gain from a play is not enormously large compared with the stake, the effect of the first of these modifications is not great. It will, however, make a difference to the situation in football pools, where a tiny stake can (very occasionally) produce huge gains—we shall return to this situation later. For the moment let us consider the effect of realizing that the game is actually unfair.

As an example, a roulette wheel may have 37 segments with the 18 even segments from 2 to 36 painted red, the 18 odd segments from 1 to 35 painted black and the last segment blank. Suppose Paul

STRATEGY IN AN UNFAIR GAME

stakes £1 on a play and each of the segments is equally likely. If he chooses red or black correctly he is given back £2, while if he chooses wrongly he loses his £1. This is the same as a simple game against the casino in which:

Probability of increasing capital by £1 is $\frac{18}{37}$
Probability of decreasing capital by £1 is $\frac{19}{37}$

We learnt in Section 6.1 that Paul can overcome the disadvantage of his limited capital by deciding in advance to restrict his potential winnings, so that he stops playing and goes home if he ever reaches his goal. So what is the effect of the changed odds on the chances of his achieving his goal? This brings us back to the random walk on the integers in which the probability of moving one step to the right is $\frac{18}{37}$, and of moving one step to the left is $\frac{19}{37}$.

As before, let q_z be the probability of ruin for Paul if he starts with z and agrees to limit his total potential capital to a. Suppose that in the game being played with unit stake, and that z grows to $z+1$ with probability p and diminishes to $z-1$ with probability q where $p+q=1$, and $p<\frac{1}{2}$ (for roulette, $p=\frac{18}{37}$). Then we get the equations:

$$q_0 = 1$$
$$q_a = 0$$
$$q_z = p \cdot q_{z+1} + q \cdot q_{z+1} \qquad (1 \leqslant z \leqslant a-1)$$

which have for their solution:

$$q_z = \frac{(q/p)^a - (q/p)^z}{(q/p)^a - 1} \qquad (z = 0, 1, 2, \ldots, a)$$

(We do not stop to explain the method which gives the solution, but the reader should check that this is the correct answer by seeing that it satisfies all the equations.)

To illustrate the effect of this small change in the odds, let us calculate the probability that Paul will be ruined at roulette before he increases an initial fortune of £70 to £80. Here $q/p = \frac{19}{18}$, $z = 70$ and $a = 80$, so:

$$q_{70} = \frac{\left(\frac{19}{18}\right)^{80} - \left(\frac{19}{18}\right)^{70}}{\left(\frac{19}{18}\right)^{80} - 1} \approx 0.48$$

This means that Paul has not much better than an even chance of increasing his capital by the modest amount of £10 before he loses the lot.

Is there any strategy Paul can adopt to improve his chances? Yes; he can vary the stakes for which he is playing. Suppose he stakes in units of £10, so that he starts with 7 units and hopes to make this into 8 units before being ruined. This means $q/p = \frac{19}{18}$, $z = 7$, $a = 8$ in the the formula, giving: q_7 = probability of ruin = 0·15, which is much better from Paul's point of view. Is this the best strategy? Not quite; in fact you can show that the best strategy for Paul to adopt is:

(i) stake £10 the first time, if he wins he has finished;
(ii) if he loses the first play, stake £20 on the second—if he wins he has won £10 more than he lost so again he has finished;
(iii) if he loses twice, then stake £40 on the third play—again if he wins he has a total of £80 and he is finished. If he loses again he is ruined.

With this strategy Paul is ruined if, and only if, he loses 3 times in succession. Hence:

$$\text{Probability of ruin} = \left(\frac{19}{37}\right)^3 \approx 0\cdot135$$

All this means that Paul has now almost made up by his cleverness for the bad odds at roulette—for if the game had been fair, the result of Section 6.1 shows that his probability of ruin would have been 0·125, and in this case it would have made no difference what strategy he adopts.

Though we have worked out mathematically Paul's best strategy, there is still something unsatisfactory about the solution. The effect of adopting this strategy is that Paul will play at most 3 times before he is either ruined or has achieved his modest goal. This would hardly occupy a whole evening, so Paul would need to be very strong-minded not to indulge further! If the object of gambling is to have a prolonged period of excitement (or misery!) then clearly the first strategy of using small stakes will achieve this—though it also makes ruin more likely. With this earlier strategy you can show that Paul should expect about 600 plays before he is finished—so that it is likely the evening would be over before he had either achieved his objective or been ruined!

Exercise 6b

To illustrate the effect of a small change in the odds, calculate in the following table the probability of ruin and success for a gambler who has a capital of z units which he wishes to increase to a units before he is ruined. At each play he stakes 1 unit and has a probability p of winning.

p	q	z	a	Probability of ruin	success
0·5	0·5	90	100		
0·45	0·55	90	100		
0·45	0·55	9	10		
0·4	0·6	90	100		
0·4	0·6	99	100		

6.3 Premium Bonds

On the grounds that it encourages savings, the British government has instituted a form of legalized gambling in which citizens are persuaded to indulge. This differs from most forms of gambling in two important respects:
  (*i*) The capital sum is never lost; only the interest which would have been earned is put at risk.
  (*ii*) The game is 'fair', as the government pays all the overhead costs and all the interest is paid out in prizes.

It is interesting to work out approximately what a citizen ought to expect when he buys a Premium Bond. Each month (after the first three) each bond has a chance of winning a prize—and a computer is employed to see that the bonds all have an equal chance. In order to make the calculation more realistic, let us suppose that George has bought 100 Premium Bonds: each month he scans the list of winning numbers hopefully. What are the chances of finding one of his bonds listed?

If we assume that there are 500 million bonds issued, on average, then we can calculate the probability of each kind of prize in a single month by dividing the total number of prizes of each kind by 5×10^6: the result is given in Table 6.1. His stake is £4·625 per year

(about 38½ new pence per month) for the government pays 4⅝%
interest on the capital, and his chances of at least one prize of any
kind in a given month is very little better than $\frac{1}{100}$. Now let us
suppose that George has no great need for his £100, so he allows it to
remain in Premium Bonds for a total of 10 years. At the end of 10
years he has still won nothing—should he feel he has been unlucky?
In terms of probability theory he has effectively gambled a stake of
£0·385 once a month—that is, 120 times, so we can use the probabilities in Table 6.1 to work out the possible outcomes.

TABLE 6.1 Probability of winning a Premium Bond prize

Value of prize	Number of prizes per month	Probability of winning at least one prize
£25 000	4	0·000 000 8
£5000	19	0·000 003 8
£1000	191	0·000 038 2
£500	191	0·000 038 2
£250	383	0·000 076 6
£100	574	0·000 115
£50	2874	0·000 575
£25	49 250	0·009 85
Total	53 486	0·010 697 6

The probability of winning at least one prize of £500 or more in
10 years turns out to be a little less than $\frac{1}{100}$ [for in a given month the
probability of not winning a prize of at least £100 is about 1 − 0·000 08
so that in 120 months the probability of never winning such a prize
is $(1 - 0·000\ 08)^{120}$, which is close to $1 - 120 \times 0·000\ 08 = 0·9904$; and
so the probability of winning at least one such prize is about 0·0096],
so there is no more than a sporting chance of that. Let us make a table
of various other possible events based on the assumption that George
holds £100 of Premium Bonds for 10 years. Their probabilities can
be calculated using the result of Example 4 in Chapter 5 (p. 88),
which gives the probability of r successes in n ($= 120$) trials—the
details of the calculations are not given since they can only be
accurately carried out using logarithms or other computing aids.
Results, given in Table 6.2, are worked out to 2 significant figures.

TABLE 6.2 Probability of winning Premium Bond prizes over a 10-year period

Event	Approximate probability
No prize of any kind	0·31
One prize	0·36
Two prizes	0·21
Three prizes	0·08
Four or more prizes	0·04
No prize of £100 or greater	0·97

George has become rather sophisticated and realizes that his money has lost value during the 10 years due to inflation. If this continues at the rate of the period 1964–69, he would need to have £154 gross at the end of 10 years if he is to purchase the goods he would have got with the original £100. This means that he will have to win at least £75 in prizes if his gamble is to pay off. We can calculate the possible results allowing for inflation, by substituting in the formula:

$$P(Z = r) = \binom{n}{r} p^r (1-p)^{n-r}$$

where $n = 120$, $r = 1, 2, 3, 4$ is the number of prizes and p is the probability obtained from Tables 6.1 or 6.2. The results are summarized in Table 6.3.

TABLE 6.3 Probabilities of effective financial gain

Event	Effect, allowing for inflation	Probability
No prize of any kind	Loss of £54	0·31
One prize of £25	Loss of £29	0·32
Two prizes of £25 or one prize of £50	Loss of £4	0·22
Three prizes of £25 or one prize of £25 and one prize of £50	Gain of £21	0·11
Prizes totalling at least £100	Gain of at least £46	0·04

The conclusion that we reach from this table is that the probability of beating inflation in a 10-year period by investing in Premium Bonds is only 0·15. Since George has become more sophisticated, he will conclude that Premium Bonds are not suitable for long-term investment. They only make economic sense as a home for money which is known to be required again in a relatively short time, say 1 or 2 years. For in such a period there is no way of beating inflation without risking large loss, and the real loss on a Premium Bond over a 2-year period is restricted to the effect of inflation, while there is a (small) chance of a substantial gain. The factor in Premium Bonds which makes them emotionally attractive is the (admittedly small) chance of a very large prize. But, if you keep £100 in bonds for a whole year, the chance of winning either a £5000 prize or a £25 000 prize are about $\frac{1}{20\,000}$; so George would have to be the one lucky person out of 20 000, each with £100 invested, if he is to get such a prize.

6.4 Football pools

During each year of the decade 1960–70, the British Public 'invested' about £100 million in football pools. Let us make an analysis of the effect of probability laws on this form of gambling. There are important differences between football pools and Premium Bonds:

(*i*) The gambler is this time staking the capital sums rather than the interest: if he does not win anything on a given Saturday, he has lost his stake.

(*ii*) The game is not fair: of all the money collected by the Pools, $33\frac{1}{3}\%$ goes to the government as tax and about 32% is retained as profits and administrative expenses so that only about 35% is available for distribution in prizes.

(*iii*) The size of the prizes is not predetermined, but depends on a very large number of factors. The total sum distributed, however, is fixed as a proportion of the takings.

In order to make realistic calculations let us assume that all prizes in the ranges shown in Table 6.4 are fixed at the value given, and that the chance of obtaining a prize from a stake of £1 is estimated by the figure in the Probability column. Although the proportion of prizes of different sizes varies from week to week the figures in the last column are a fair estimate over a prolonged period based on the prizes actually given by one of the large pools firms (you obtain the probability of a prize in, say, the range £3000–£10 000 by dividing the total number of such prizes in a season by the total stake money). The sum that would be paid out each week if our model were exact would be about 35% of the stake money—which is the situation in real life—so let us calculate the expected outcome using our model.

TABLE 6.4 Probability of winning a football pool prize

Range	Fixed prize	Probability
More than £75 000	£100 000	0.5×10^{-6}
£25 000–£75 000	£50 000	10^{-6}
£10 000–£25 000	£15 000	2×10^{-6}
£3000–£10 000	£5000	8×10^{-6}
£1000– £3000	£2000	2×10^{-5}
£300– £1000	£500	7×10^{-5}
£50– £300	£100	10^{-4}
£5– £50	£20	5×10^{-3}

We would expect the real-life situation to correspond only approximately to the results we get from the model, but the real-life pools are no more favourable than our model.

In order to get realistic figures, let us assume that George spends £1 per week on football pools for 40 weeks in the year regularly during the 40 years of his working life. During this period he gets 5 prizes of £20 and one of £500: should he feel that he has been unlucky? He was really hoping for a prize of £100 000. Let us this time ignore inflation, as it affects both the stake money and the prize money equally. We calculate the probabilities of certain interesting events. The exact probabilities of events 1–10 in Table 6.5 could be found in this case by using the result of Example 4 of Chapter 5 (p. 88) giving the probability of r successes in n ($= 1600$) trials— but the calculation would be extremely laborious. There are various approximation methods available which help if, as here, p is very small and n is very large. Events 11 and 12 have probabilities estimated from events 1–10.

TABLE 6.5 Probabilities of winning football pool prizes over a 40-year span

Event	Probability
1. No prizes of any kind	0·0004
2. Between 1 and 3 prizes won	0·05
3. Between 4 and 11 prizes won	0·81
4. Between 12 and 15 prizes won	0·12
5. More than 15 prizes won	0·06
6. No prizes £500 or more	0·83
7. One prize £500 or more	0·16
8. Two or more prizes of £500 or more	0·01
9. Two or more prizes of £100	0·01
10. At least one prize of £50 000 or more	0·0025
11. Total winnings of £1600 or more	0·09
12. Total winnings of £500 or more	0·24

This shows that regular betting on the pools results in an almost certain loss over a lifetime. In 40 years George would have spent £1600 and his chances of receiving that amount back in prizes is less than $\frac{1}{10}$. In fact, his chance of even receiving as much as £500 back in prizes is less than $\frac{1}{4}$. It is not the situation where George gets ruined, since presumably he receives wages each week and simply chooses to use part of them in this way.

For comparison purposes it is worth calculating the result of saving £40 per year regularly through a working life of 40 years. For example, one could purchase a life insurance policy for the 40-year term for a capital sum of £1850, with profits payable. Assuming the average rate of profits in the period 1964–69 (if one avoids the industrial life insurance companies who collect door to door and therefore have high expenses, a good insurance policy in this period yielded about £4% pa allowing for special bonuses), this would yield about £6000 if George survived the 40 years. Again if we allow for inflation, his investment of £1600 over the years would require to be worth about £4750 to allow for rising prices; so that he would expect to make a profit on his policy of about £1250 in real terms. If George did not survive the 40 years, his heirs would receive a somewhat larger proportionate profit. Summarizing the two ways that George could regularly 'invest' part of his earnings, we have

(i) £40 a year on football pools: George will lose overall with a probability $\frac{9}{10}$; he will lose more than £1000, with a probability $\frac{3}{4}$.

(ii) £40 a year as a premium on a life insurance policy: George (or his heirs) is practically certain to gain overall. At current rates of profits and inflation, he will gain more than £1000.

There can be little doubt which course of action makes sense for George!

6.5 Expectation

During this Chapter we have been making various calculations, and we have tried to distinguish between games which are fair and games which are unfair. Our next objective is to make this notion precise, and at the same time to give it a numerical value.

Suppose George plays a rather complicated game for which the outcome depends on probability laws—there are various possible results:

e_1 which happens with probability p_1
e_2 which happens with probability p_2
.   .   .
.   .   .
.   .   .
e_k which happens with probability p_k

This means the probability space S has k points in it $e_1, e_2, \ldots e_k$.

Now suppose for each result e_1, allowing for the stake money, there is a nett payoff $\chi(e_1)$, calculated by subtracting the stake from the prize received. [$\chi(e_1)$ may be positive for some e_1, zero for others and negative for others.]

Now consider what happens when George repeats the game a large number of times under conditions of independence. In a large number, n, of trials:

e_1 occurs r_1 times giving a payoff $r_1\chi(e_1)$
e_2 occurs r_2 times giving a payoff $r_2\chi(e_2)$
.   .   .
.   .   .
.   .   .
e_k occurs r_k times giving a payoff $r_k\chi(e_k)$

Adding, we obtain the total gain from n repetitions:

$$r_1\chi(e_1) + r_2\chi(e_2) + \ldots + r_k\chi(e_k)$$

so the average gain per game is:

$$\frac{r_1}{n}\chi(e_1) + \frac{r_2}{n}\chi(e_2) + \ldots + \frac{r_k}{n}\chi(e_k).$$

The frequency interpretation of probability which we discussed in Chapter 5 shows that r_i/n is close to p_i when n is large, so that the average gain per game gets close to:

$$p_1\chi(e_1) + p_2\chi(e_2) + \ldots + p_k\chi(e_k) = E(\chi)$$

as the number of games gets larger and larger. The number $E(\chi)$ is called the *expectation* of gain from playing the game. If it is positive, then the game favours the player; if it is negative, the game is against the player.

A game is said to be *fair* if $E(\chi) = 0$. Otherwise it is unfair.

Example 1. In the game between Peter and Paul discussed in Section 6.1 there are two outcomes, e_1, e_2, each with probability $\frac{1}{2}$. This gives:

$$\chi(e_1) = 1$$
$$\chi(e_2) = -1$$
and
$$E(\chi) = \tfrac{1}{2}\chi(e_1) + \tfrac{1}{2}\chi(e_2)$$
$$= \tfrac{1}{2} - \tfrac{1}{2} = 0$$

Our intuition that this was a fair game agrees with the definition.

Example 2. In the game of roulette, as discussed in Section 6.2 there are 37 outcomes, $e_1, e_2, \ldots e_{37}$ each with probability $\frac{1}{37}$. If Paul bets on black, then:

$$\chi(e_1) = \chi(e_3) = \ldots = \chi(e_{35}) = 1$$
$$\chi(e_2) = \chi(e_3) = \ldots = \chi(e_{36}) = -1 = \chi(e_{37})$$

In this case $E(\chi) = -\frac{1}{37}$ since there are 19 results with a payoff of -1 and 18 results with a payoff of $+1$. The game is not fair, and in a large number n of trials we would expect Paul to lose about $n/37$ if he made a unit stake each time.

Exercise 6c

1. Table 6.1 indicates the probabilities of possible results of the draw for prizes in Premium Bonds. If G is the prize money received, calculate $E(G)$ and show that it is equal to the stake (which we saw was £0·385). This makes it a fair game.
2. Carry out the same exercise for football pools using the model assumed in Table 6.4. Check that this game is not fair by comparing the expected gain with the stake of £1.

6.6 Wisdom in gambling

We have given the impression in this Chapter that gambling in an unfair game is foolish. We should point out that there are situations in life where it is prudent to gamble in this way. This is the principle of insurance. If one insures the house one owns against fire then a small premium is paid each year and most people never get anything back. There is a small chance that the house will burn down, and in this case, the owner receives from the insurance company a sum of money to repair it or replace it. Let us work out the probability situation for this case. George has a house worth £8000 which he insures by paying £10 pa to Transfire United. In one year, any amount can be claimed, but let us again simplify the model so that there are just 5 possible outcomes (see Table 6.6).

TABLE 6.6 Outcomes of house insurance

	Event	Probability	χ
e_1	No claim during year	0·89	−10
e_2	Total loss, claim £8000	10^{-4}	7990
e_3	Substantial fire, claim £4000	10^{-4}	3990
e_4	Bad burglary, claim £100	10^{-2}	90
e_5	Broken window, claim £10	$\frac{1}{10}$	0

Again, the real-life situation is much more complicated as the amount of claim can be any figure corresponding to the loss—our simplified model gives a fair approximation to what happens in practice.

116   THE GAMBLER'S RUIN

This gives $E(\chi) = -6 \cdot 8$, so that the technical expectation of net loss from the stake of £10 is more than half of it. One could carry out the sort of analysis in Section 6.4 to compute the likely result over a lifetime of paying house insurance, and it is obvious that, for most people, it is not a good investment in money terms. There are heavy administrative expenses involved in collecting premiums and paying claims, and the insurance companies have to make a profit on average —so it should not surprise us that our technical expectation of gain is negative. In spite of this, it is a good idea to insure your house, for you are taking steps to preserve what is probably your largest investment against the (admittedly small) chance of total loss.

The net gain χ which we have been using is an example of what mathematicians call a *random variable*. Technically χ was defined for each possible outcome so it is a function $\chi : S \to R$ on the probability space. We say that $E(\chi)$ is the expectation of the random variable χ.

Example 3. At the village fair there is often a game which consists of rolling a penny down a slope onto a board marked with numbered squares. If the penny lands completely inside a square, the house pays

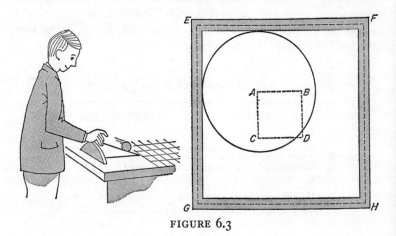

FIGURE 6.3

the number of pennies marked, if the penny touches or falls across a boundary line, then it is lost.

Imaginary dotted lines in the middle of the thick boundary lines of the squares marked on the board will form a grid of equal squares. A coin can end up anywhere on the board, but its centre will always lie in one of the dotted squares, and all positions for the centre of the coin inside any dotted square are equally likely. For a win, the coin must end up inside the squares marked on the board, and for this to happen, the centre of the coin must end up in the smaller dotted square $ABCD$ (see Fig. 6.3). Hence the probability of a win is the ratio:

$$\frac{\text{area of square } ABCD}{\text{area of larger dotted square}}$$

This may differ for different fairgrounds, but a typical value is about $\frac{1}{10}$. Since the payoff varies usually from 2p to 10p according to the square, it is clear that this game is unfair to the player. In fact, the expectation of net gain $E(\chi)$ is likely to be worse than $-\frac{1}{2}$p for a stake of a penny.

Exercise 6d

The outcome of throwing a symmetric die is a digit 1, 2, 3, 4, 5 or 6 each with probability $\frac{1}{6}$. If you received a payoff equal to the amount of the digit, what should the stake be to make it a fair game?

7 | TOPOLOGY, OR THE SHAPE OF A SET

7.1 The nature of mathematics

Geometry was probably the first subject which was studied rigorously. However, there was not much progress in the subject from the time of the Greeks until the nineteenth century. About the middle of the nineteenth century, there was a completely new development in geometry which has been one of the great forces in the growth of the mathematics of the twentieth century. The new idea was to study the properties of geometrical figures which persist even when the figures are deformed so drastically that the metric properties (*i.e.*, the properties depending on distance) are completely lost.

The subject started in Germany, mainly at Göttingen. A bright student called Riemann went there to study in 1845—he found that the mathematical ideas current in Göttingen had already been greatly affected by the new geometry. Later Riemann was to be one of the world's great mathematicians, who laid the foundations of modern complex function theory. In this subject the ideas of what is called topology give a depth of understanding that would not have been possible otherwise, and Riemann's effort at that time provided an impetus to the later development of topology.

The subject of topology has become a central focus for mathematical thinking today. Here the ideas of algebraic structure and analysis meet. One of the rewarding aspects of mathematics is the extent to which ideas developed in one area turn out to be useful in apparently disconnected fields of study. Sometimes, as in the theory of sets, the unifying idea turns out to be a structure underlying all mathematical thought. In the main, topology unifies because it uses and brings together ideas from many different fields. Of course, having found a use for an idea, it frequently happens that not enough is known about its detailed structure—and this then provides a stimulus to research.

The process we have just described explains in large measure the nature of research in mathematics. There has been an explosion of mathematical ideas since 1950, so that, at the time of writing (1970) about 15 000 papers per year are written either developing some new idea in a branch of mathematics or exploring the interrelationships between different branches of the subject, or seeking to solve a problem suggested by one of these relationships. Research in mathematics is creative, and it is very exciting for those involved in it. Earlier in the history of mathematics, applications to the physical world provided an important stimulus to research. Nowadays this is less important, and applications to other branches of mathematics are more likely to provide a growth point.

Enough of generalities—let us get back to the subject for this Chapter. Because topology relies on deep results from a number of different fields, it is not possible to go far in a book of this kind in presenting ideas of topology. However, we will try to introduce the subject by discussing some of the important intuitive concepts. The first topic we turn to is polyhedra. These were studied by the Greeks, but they did not notice an important formula relating the numbers of edges, vertices and faces of any simple polyhedron.

7.2 Euler's formula for polyhedra

By a polyhedron we simply mean a solid whose surface consists of

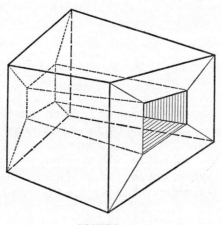

FIGURE 7.1

plane faces each in the shape of a polygon. It is said to be *simple* if it has no 'holes' in it—so that its surface, if made of rubber, could be continuously deformed into a sphere. Figure 7.1 illustrates a block with a rectangular hole cut out of it. This is not a simple polyhedron.

The Greek geometers knew of five regular polyhedron (simple polyhedron such that the faces are all congruent, *i.e.*, if we cut out a sheet of paper the exact size and shape of one face it could be fitted exactly over each of the faces, and have their angles all equal). These are illustrated in Fig. 7.2, and we shall prove in Example 1 that there are no others.

For any simple polyhedron, let us denote by v the number of vertices, e the number of edges and f the number of faces. Then it was noticed by Descartes as early as 1640, and rediscovered by Euler in 1752 that, no matter how complicated the figure:

$$v - e + f = 2$$

is always true.

Exercise 7a

Count v, e, f for each of the regular solids in Figure 7.2 and check that the formula holds. Draw a non-regular simple polyhedron of any shape and check the formula. Does the result hold for the solid of Fig. 7.1?

Proof of Euler's formula

Let us give an intuitive proof of this formula. Imagine the simple polyhedron is hollow with a surface made of rubber. Cut out one of the faces and remove it—this will reduce the number of faces by 1, but will not change the number of edges or vertices. Deform the remaining surface until it is flat on the plane, but is not overlapping in any way. In doing this operation, the areas and angles are bound to be altered, but you end up with a network of non-overlapping polygons in the plane, one for each of the remaining faces of the polyhedron. It is therefore sufficient to show that for any such plane network:

$$v - e + f = 1$$

EULER'S FORMULA FOR POLYHEDRA

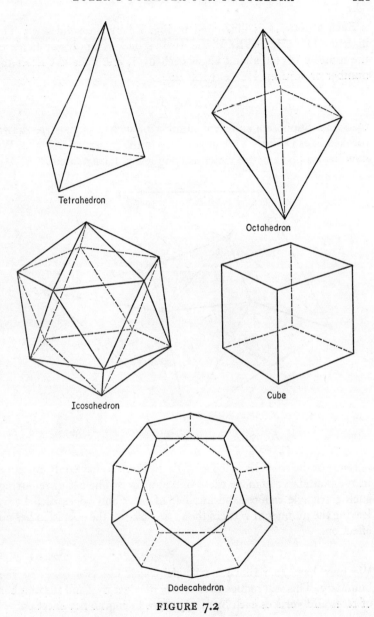

FIGURE 7.2

Such a network is illustrated in Fig. 7.3 by the solid lines. The insertion of a diagonal by joining two vertices of a polygon increases the number of 'faces' and edges each by 1, and does not affect the number of vertices. Hence it leaves:

$$v - e + f$$

the same. Suppose a sufficient number of such diagonals are drawn (broken lines in Fig. 7.3) to divide each of the faces into triangles. We now have a plane figure which is completely 'triangulated'.

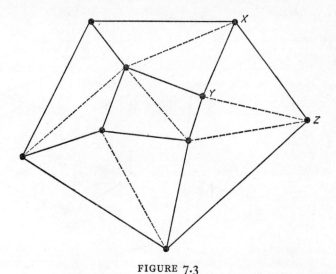

FIGURE 7.3

Some of the triangles (such as XYZ in Fig. 7.3) have only one edge in the boundary. Remove these one at a time. The act of removing such a triangle reduces the number of faces and edges each by 1, leaving the number of vertices the same, so again the operation has no effect on:

$$v - e + f.$$

We may have to remove a triangle which has two edges in the boundary. This will reduce the number of edges by 2 and the number of faces and vertices each by 1; again producing no net effect on:

$$v - e + f.$$

After a finite number of steps we are left with a single triangle with 3 vertices, 3 edges and 1 face, for which:
$$v - e + f = 1.$$
It follows that $v - e + f = 1$ for any polygonal network so that for any polyhedron we must have:
$$v - e + f = 2.$$

Example 1. Show there are only 5 regular polyhedra. For suppose there are f faces, each face having n vertices and n edges. Then since each edge of the polyhedron is an edge of 2 faces meeting there:
$$nf = 2e$$
where e is, as before, the total number of edges of the polyhedron. Now suppose r edges meet at each vertex. Then each edge ends in 2 vertices so that:
$$rv = 2e$$
where v is the total number of vertices of the polyhedron. Substituting for f and v in Euler's formula gives:
$$\frac{2e}{n} + \frac{2e}{r} - e = 2$$
or
$$\frac{1}{n} + \frac{1}{r} = \frac{1}{2} + \frac{1}{e} \tag{7.1}$$
We know that $n \geqslant 3$, $r \geqslant 3$ since polygons have at least 3 sides, and at least 3 edges meet at a vertex. It is not possible to have $n \geqslant 4$ and $r \geqslant 4$ for then the left hand side of equations (7.1) would be $\leqslant \frac{1}{2}$, which is impossible. Hence there are two cases:

(i) $n = 3$, giving:
$$\frac{1}{r} = \frac{1}{6} + \frac{1}{e}$$

with $r = 3, 4, 5$ being the only possible values of r, since the left-hand side must be $> \frac{1}{6}$. These values result in the tetrahedron, octahedron, and icosahedron (for which $e = 30, f = 20, v = 12$).

(ii) $r = 3$ similarly gives $n = 3, 4$ or 5 as possible values corresponding to the tetrahedron, cube and dodecahedron (for which $e = 30, f = 12, v = 20$).

The five regular solids are illustrated in Fig. 7.2.

Exercise 7b

The Euler formula for a network of polygons in the plane can be applied to any map. Consider the map of the mainland of Wales (Fig. 7.4). Mark on it the points where at least 3 counties meet: these are vertices. The county boundaries between two vertices are edges, and the counties are faces. Check that $v - e + f = 1$.

FIGURE 7.4

7.3 Topological properties

The Euler formula we have just been considering would apply equally well to any subdivision of the surface of a sphere into regions bounded by curved arcs. If we imagined the surface of the sphere made of a flexible material like rubber, the formula will still hold if the surface is deformed into any shape as long as the rubber is not torn or folded over on itself. The relationship between edges, vertices and faces is an example of a topological property, for it is so fundamental that it is not changed by a large class of transformations.

Let us try to make our notion of a topological property more

TOPOLOGICAL PROPERTIES

precise. For any two sets A, B in space, or on the plane a function $f: A \to B$ is said to be a topological transformation of A onto B provided:

(i) the function of f is (1, 1) and onto:

(ii) both the function f and its inverse $(f^{-1}: B \to A)$ are continuous, which means that for each fixed point $p_0 \in A$, we can arrange that $q = f(p)$ is as close as we wish to $q_0 = f(p_0)$ by making p close to p_0, and conversely we can ensure that p is close to p_0 by making q close to q_0.

For example, a map of a country or a continent such as is used in geography is a topological transformation of that part of the earth's surface onto a part of the plane sheet of paper. But note there is no way of constructing a map of the whole globe on a plane piece of paper so that it is a topological transformation. There is a fundamental difference between the complete surface of a sphere and a plane. However, if you remove a single point, say the North Pole, from the sphere, then the remainder can be mapped by a topological transformation onto a plane. For example, Fig. 7.5 illustrates the mapping

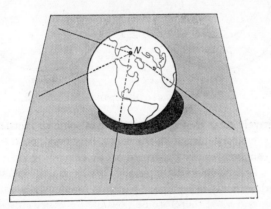

FIGURE 7.5

obtained by projecting from N, the North Pole, onto a tangent plane through S, the South Pole. (Note that this is not a form of mapping which would appeal to a geographer: why not?)

Any property of a geometrical figure which remains true for every figure into which it may be mapped by a topological transformation is called a *topological property* of A. If A, B are two figures such that there is at least one topological transformation of mapping A onto B, then the two figures are *topologically equivalent*.

Example 2. A torus is the surface of a ring doughnut, or a bicycle tube without the valve. If you think of a torus made out of Plasticine, it could be pushed around or deformed into the shape of Fig. 7.1. Hence this polyhedron is topologically equivalent to a torus. Every simple polyhedron is topologically equivalent to a sphere. But note that the torus and sphere are not equivalent—to see this we have to think of some property which is preserved by topological transformations, and which is held by a sphere but not by a torus. Consider a simple closed curve on the surface of a sphere—that is a curve which does not cross itself and comes back to its starting point (see Fig. 7.6).

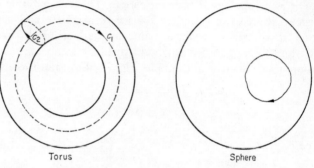

Torus Sphere

FIGURE 7.6

It is intuitively clear that such a curve has some properties not held by all simple closed curves on the torus. Any curve on the surface of a sphere can be steadily deformed while remaining on the surface until it has shrunk to a single point. This shrinking operation could be mapped onto any other figure which is topologically equivalent to the sphere, and must therefore be possible for every closed curve on such a surface. But, in Fig. 7.6 it is clear that neither of the curves c_1, c_2 can be shrunk to a point while remaining in the surface of a torus.

Exercise 7c

Draw two closed curves on the surface of the polyhedron of Fig. 7.1, neither of which can be shrunk steadily to a point.

TOPOLOGICAL PROPERTIES

Example 3. A country consisting of two islands is not topologically equivalent to a country consisting of a single island. For given any two points p, q on a single island one can get from p to q by a path lying inside the country (see Fig. 7.7). If country B were topologically

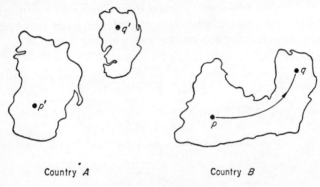

Country A Country B

FIGURE 7.7

equivalent to country A, this path would map into a path joining two points in A. This is impossible if p, q are in different islands. Mathematicians call a set E *pathwise* (or arcwise) *connected* if, for each pair p, q of points in E, there is a curve joining p to q which lies completely inside E. This is a topological property.

Example 4. In the plane, not all connected sets are topologically equivalent. Figure 7.8 illustrates 3 sets in the plane; each clearly has

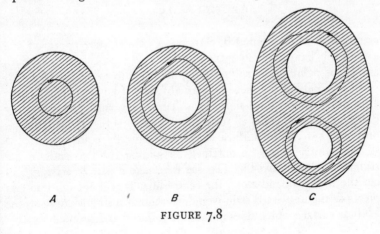

A B C

FIGURE 7.8

this property of being pathwise connected. The set A has the property that any simple closed curve in A can be shrunk to a point, whereas B and C do not have this property. We say that A is *simply connected*, and B and C are *multiply connected*. B and C are not equivalent either, for in C there are two distinct curves which cannot be deformed into each other or to a point, but in B any simple curve not deformable to a point could be deformed into the curve shown.

Exercise 7d

Take a long rectangular piece of paper, and paste the two ends together after giving one of them a half-twist (through 180°), as in Fig. 7.9. This surface is called a Möbius band—after the German

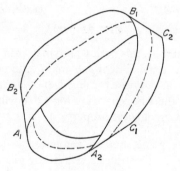

FIGURE 7.9

astronomer who invented it. Suppose a painter starts work on one side of this band (say along the piece from A_1, to A_2 in the diagram) and keeps painting, he will go along the back up to B_1, then from B_1 to B_2 and behind the piece he started from. If he keeps going, then after going round once more he will have returned to the starting point and have covered all the strip. This means that the surface has only one side. This surface is not topologically equivalent to the surface which would be obtained by joining the two ends of the rectangle without twisting. To see this, take a pair of scissors and cut the Möbius band along the centre line. It will not fall into two pieces as the untwisted strip would, so there is a simple curve on the Möbius surface which does not divide it into two disjointed regions.

Example 5. The Möbius band is an example of a one-sided surface with edges. Since most of the surfaces we can think of that have no edges (*e.g.*, sphere or torus) have two sides, it is interesting to ask whether we can define a closed surface which has only one side. Technically, we say that a surface is closed if each of its points is interior; that is, there is a small circle (or approximate circle) on the surface with that point as centre which contains only points of the surface. The first such one-sided closed surface was devised by a German mathematician, Felix Klein, and is therefore called Klein's bottle. It is illustrated in Fig. 7.10. Think of this as a horn shape in

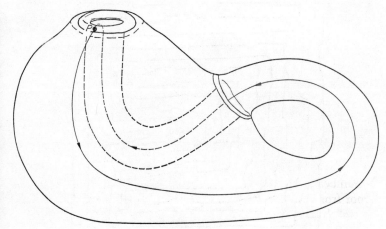

FIGURE 7.10

which the thin end is drawn through the surface near the base and brought level with the base then being connected to the base by a plane annulus (shape *B* in Fig. 7.8). Imagine a fly walking from P on the path shown. It will arrive on what appears to be the underneath of the annulus and so can get to 'the other side' of the surface at P. This means that the surface has only one side—and the bottle has no inside or outside! Thus, Klein's bottle is not topologically equivalent to either of the surfaces in Fig. 7.6.

7.4 Colouring a map

We have already been assuming as intuitively obvious the idea that any simple closed curve in the plane divides the plane into two sets, the inside set *I* and the outside set *O*. Here each of *I* and *O* are pathwise connected by paths not crossing the original curve, but it is impossible to find a path joining a point of *I* to a point of *O* which does not cross the curve. This result is called the *Jordan curve theorem*, and it seems a very obvious result for curves such as a circle, a rectangle or an ellipse. However, a simple curve may be extremely complicated: it may not even be possible to draw it, for it could have infinite length, but Fig. 7.11 illustrates a curve where it is

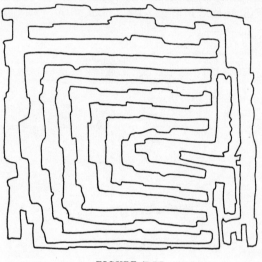

FIGURE 7.11

not immediately obvious which points are inside and which are outside. This theorem is a very deep one, and a rigorous proof is not at all easy—in fact Jordan's original proof turned out to be invalid! It is reasonable to think that one could find out whether a point P is inside or outside by drawing a line through P and counting how many times this line cuts the curve starting from P until the line is clearly outside. If the number is even then P is outside, whereas an odd number means that P is inside. Unfortunately we cannot make a proof from this idea, partly because the number of intersections may well be infinite—and we cannot distinguish between an even and an odd infinity!

Now assume that the Jordan curve theorem is valid and suppose

COLOURING A MAP

we have a geographical map in the plane of a number of countries each of which is the interior of some simple closed curve. In geography it is customary to give different colours to any two countries which have a piece of common boundary. The fewer colours used the cheaper it will be to print the map. If the map is complicated what is the minimum number of colours required? Figure 7.12 is a map of an

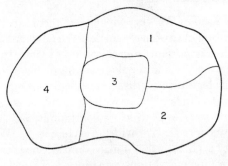

FIGURE 7.12

island containing four countries each of which touches the other three. This map would require four colours.

Exercise 7e

Draw a complicated fictitious map and indicate (by numbering countries with the same number if they can be coloured the same colour) how it could be coloured economically. No matter what map you start with you ought to be able to colour it with 4 colours.

Your experience should suggest the following:

> For any subdivision of the plane by simple curves into non-overlapping regions, it is always possible to mark the regions with one of the numbers 1, 2, 3, 4 in such a way that no two contiguous regions have the same number.

No map requiring more than four colours is known, and yet the combined efforts of many famous mathematicians have not produced a proof of this conjecture—though many false 'proofs' have been given! The theorem has been proved true for any map containing not more than 37 regions.

One curious feature of this problem is that its solution is possible (and not too difficult) in more complicated circumstances. For example, on the surface of a torus it has been proved that any map can be coloured using seven colours, and there is a map of seven regions each of which touches the other six.

7.5 Fixed-point theorem

We said, in Section 7.1, that topological theorems turn out to be useful in other branches of mathematics. We now outline a proof of a theorem which is important from this point of view. It illustrates the fact that not all theorems in topology are immediately obvious to the intuition. Consider a disc D, that is, the region of a plane consisting of the interior of a circle together with its circumference. Suppose we have a mapping $f: D \to D$ which is continuous, i.e., for each point p_0, $f(p)$ will be close to $f(p_0)$ provided p is close to p_0. [We do not need to assume that f is (1, 1) or that it maps onto D, so we are not restricting attention to topological transformations f, though the theorem is true for such transformations]. If we think of D as a rubber disc, then it could be stretched, folded, shrunk or deformed in any way other than tearing, provided the final position of each point was inside the original set D. The mathematician, Brouwer, proved Theorem I.

THEOREM I

Any such transformation f leaves at least one point fixed.

This means that there is at least one point p_0 such that $f(p_0) = p_0$. Note that it is obvious that there may not be more than one fixed point, for f could be simply a rotation of the disc D through 90°.

We can prove this theorem by the *reductio ad absurdum* method of proof—by assuming the theorem is false and deducing a contradiction. Suppose then that $f(p)$ is different from p for all points p in D. This means that for each point p in D we can draw the arrow from p to $p' = f(p)$ to denote the motion carried out by the mapping f. Consider what happens to the boundary points of the disc. All the corresponding arrows have to point into the circle, since p' is always in D. (Note that we are considering a mapping $f: D \to D$ from a set to itself so you should think of the range set D as superimposed on the domain set D.)

FIXED-POINT THEOREM

Suppose we begin at point p_1 on the circumference and consider all the points on the circumference in anti-clockwise order until we return to p_1 (see Fig. 7.13). For the sake of simplicity we show 16 points. Since a small change in position of p on the circumference (*i.e.*, transferring our attention from point 1 to a point very near to it) results in a small change in the position of $f(p)$ (for the image of any point near 1 has to be near to the image of 1), the transformation arrow (which starts on the circumference) must change direction

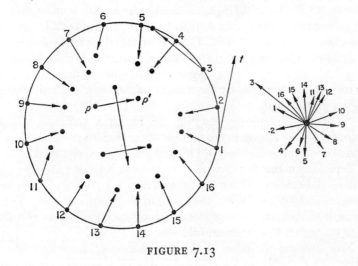

FIGURE 7.13

smoothly as we move to neighbouring points on the circumference. In other words, the direction of the arrow is only slightly different for points on the circumference which are very near to each other. When p has gone round completely to its starting position p_1, the final position of the arrow is the same as the starting position, for this arrow joins p_1 to $f(p_1)$. Hence the total angle turned through by these transformation arrows must be an integer multiple of 360°. The first step is to prove that it must be exactly 360°.

Consider the tangent arrow t at the moving point on the circumference. As we describe the circle once, this arrow clearly does turn through 360°. If the total angle turned through by the transformation arrow is different from 360° then the direction of the transformation arrow must cross that of the tangent arrow at least once as p moves round the circumference. But both arrows turn continuously—so by the property discussed in Section 4.5, there must be some point where the transformation arrow points along the tangent. This is impossible as it always points inwards. Figure 7.13 illustrates that the movement of the transformation arrow as p moves through 16 points on the circumference is through one complete revolution.

Now repeat this argument on other circles inside the disc with the same centre as D. If we consider the arrow starting from the point corresponding to p on a neighbouring circle, this involves a small movement in p, so that $f(p)$ again changes by a small amount, and the transformation arrow from p to $f(p)$ will change (in both length and direction) by a small amount. This means that if we let p move round a circle of radius just less than 1, and construct a diagram similar to Fig. 7.13, the change in the diagram will be small so that the number of revolutions described by the transformation arrow can only change by a small amount.

So if we arrange (by varying the radius a sufficiently small amount) for the change in the number of revolutions to be less than $\frac{1}{2}$, there cannot be any change. (This argument is a very useful trick—it depends on the fact that if n, m are unequal integers then the difference between n and m has to be at least 1, so that if n, m are two integers and $-\frac{1}{2} < n - m < \frac{1}{2}$, then we can deduce that $n = m$.) Hence on each of the concentric circles the transformation arrows will also turn through precisely one revolution.

Now consider the effect of the mapping on O, the centre of the circle. This again gives an arrow OA with a definite length and direction, and all points p close to O will have transformation arrows whose direction is close to OA. If p describes a small enough circle centred at O, it follows that all the transformation arrows point more or less in the same direction, so if we construct the second diagram in Fig. 7.13 for this small circle, the transformation arrow must describe zero revolutions. This contradicts our previous conclusion that the transformation arrow turns through one revolution on every circle centred at O. Our initial assumption that the theorem is false has led us to a logical contradiction—so the theorem must be true.

Exercise 7f

A section of a curve such as AB in Fig. 7.14 may be called an *arc*, and the ends of the arc are called *nodes*. When arcs are combined we

FIXED-POINT THEOREM

say we have a *network*. We say that a node is even if there is an even number of arcs which enter it—otherwise it is odd. Thus A and C are odd, whereas B and D are even in Fig. 7.14.

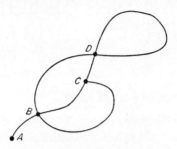

FIGURE 7.14

Now start doodling! Show that it is possible to draw a network without lifting the pencil or covering any arc twice if and only if there are no more than 2 odd nodes. Check with Fig. 7.15. Where

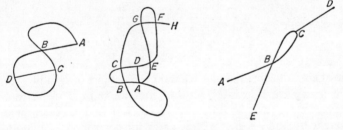

FIGURE 7.15

must you begin and end if there are two odd nodes? Use this information to design treasure hunts or circuits.

8 | PARADOXES OF THE INFINITE

8.1 A whole equal to part of itself

In Chapter 3 we discussed the set N of whole numbers or positive integers. These are the numbers we use for counting: suppose we try to 'count' the even integers 2, 4, 6, ... We could do this in the usual order by 'picking up' mentally the number 2 and saying 'one', then picking up the number 4 and saying 'two', and so on. This means that we are setting up a correspondence f from the set E of even integers to the set N of counting numbers, as shown in Fig. 8.1. This is a

$$f: \begin{matrix} 2, 4, 6, 8, \ldots & , 2n, \ldots \\ \updownarrow \updownarrow \updownarrow \updownarrow & \updownarrow \\ 1, 2, 3, 4, \ldots & , n, \ldots \end{matrix}$$

FIGURE 8.1

special kind of function—for we want each object in E to be counted (that is, assigned an integer $r \in N$) and we want it to be counted precisely once (that is, different elements of E must correspond to different numbers r in N). Now every integer r in N is going to be used in this counting for the even integer $2r$ maps to r. Thus the function $f : E \to N$ is (1, 1) and onto.

If the process of counting is valid for the set E, what we have shown is that there is the same 'number' of even integers as there are integers. But E is a proper subset of N. In fact if we consider the complement $D = N - E$, consisting of the odd integers, this also has the same number of elements as N by using the correspondence $g : D \to N$ illustrated by Fig. 8.2. This function is again (1, 1) and

$$g: \begin{matrix} 1, 3, 5, 7, \ldots & , 2n-1, \ldots \\ \updownarrow \updownarrow \updownarrow \updownarrow & \updownarrow \\ 1, 2, 3, 4, \ldots & , n, \ldots \end{matrix}$$

FIGURE 8.2

onto. This means that we have split the set N into two disjoint subsets, D and E, each of which has the same number of elements as N itself.

Some famous mathematicians through the ages have been puzzled by this difficulty (see Fig. 8.3). They avoided it without understanding it, and simply decided that it was not legitimate to compare the

FIGURE 8.3

numerical size of sets when these were 'not finite'. The solution to the paradox was provided finally by George Cantor, a brilliant young German mathematician, who was born in 1873. Our object in this Chapter will be to understand the basic ideas behind Cantor's reasoning—ideas which are extremely simple.

The first step is to understand more clearly the process of counting for finite sets. We learn, very young in life, that if we are counting, say, a box of bricks, it does not matter how many times we do it, or how many different people do it, the answer should always be the same. If John and George both count the bricks and John says there are 47, George says there are 49, what do we conclude? An intelligent creature who had never thought about counting might, in this situation, decide that the counting process did not lead to a unique answer. We know better, and conclude that either John or George (or both) made a mistake since there has to be a unique whole number n such that the box contains n bricks. This intuition we have about the counting process corresponds to a deep property of the set N of positive integers.

Suppose the bricks in the box are counted twice, and on each occasion the counting is done without a mistake. This means that we have two functions f, g each defined on the set B of bricks (Fig. 8.4):

$$f : B \to \{1, 2, 3, \ldots, n\}$$
$$g : B \to \{1, 2, 3, \ldots, m\}$$

and each function is (1, 1) and onto an initial segment of the positive integers. How do we know that the last numbers n, m in these segments must be the same? Since f is (1, 1) and onto, it has a unique

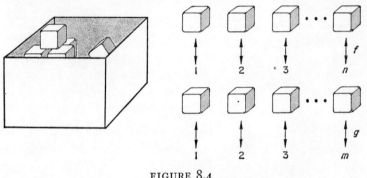

FIGURE 8.4

inverse function $h : \{1, 2, \ldots, n\} \to B$ (see Section 2.4). Now take the composition function $g \circ h$ which maps $\{1, 2, \ldots, n\}$ onto $\{1, 2, \ldots, m\}$:

$$\{1, 2, \ldots, n\} \stackrel{h}{\to} B \stackrel{g}{\to} \{1, 2, \ldots, m\}$$

$$g \circ h$$

This mapping is also (1, 1), which leads to a contradiction unless $n = m$. So our assumption that the result of counting a box of bricks leads to a unique whole number n is equivalent to Theorem I.

THEOREM I

For any positive integer, n let E_n denote the set of positive integers not greater than n. Then, if $n \neq m$, it is not possible to find a (1, 1) correspondence between E_n and E_m.

This theorem can be proved (the proof is not trivial) using the properties which we are assuming of the set N of positive integers. Thus we are reassured in our belief that counting works at least for a box of bricks—indeed for any finite set. The sort of paradox with which we started this Section is a paradox precisely because it cannot happen when we are counting a finite set, however large.

Example 1. In a certain classroom there are 23 desks each capable of accommodating precisely one pupil. If the teacher comes in and finds that every desk is occupied he knows that there are 23 pupils in his class. If, however, he sees that some desks are empty, he knows that fewer than 23 pupils are present, and, if all the desks are full and some people are standing, he knows there are more than 23 pupils present. The process of seating the class in the room sets up a (1, 1) correspondence between pupils seated and desks occupied, so that we can deduce facts about relative numbers of pupils and desks by observing either pupils standing, or empty desks.

Exercise 8a

1. In a certain village there are 40 wealthy people, 400 poor people, 100 wise people and 10 people who are both wealthy and wise. Assuming everyone in the village is either wealthy or poor (not both) and either wise or stupid (not both), how many stupid people are there? How many people are wealthy and stupid?
2. A set A has 20 members, a set B has 100 members, and their union has 100 members. What can you say about the sets A, B?
3. Find all the (1, 1) correspondences between the sets $\{1, 2, 3\}$ and $\{1, 2, 3\}$
*4. Show that there are $n!$ different (1, 1) correspondences between the set $\{1, 2, \ldots, n\}$ and itself. (Remember that $n!$ stands for the product of the integers $1, 2, \ldots, n$.)

8.2 Cardinal number

In Example 1 the actual number of desks in the room was not important. The number of pupils was greater than the number of desks if some pupils were left standing, and the reverse was true if some desks remained empty. If every desk was filled and no pupils were left over then we deduced that there was the same number of desks as pupils. This is the germ of Cantor's great idea.

Two sets A, B have the same *cardinal number* if there is a mapping $f: A \to B$ which is (1, 1) and onto. We use the notation:

$$A \sim B$$

for the statement that A and B have the same cardinal. This defines a relation $\sim$ between sets, which clearly has three properties which we illustrate in Fig. 8.5 for finite sets, though the properties are valid also for infinite sets:

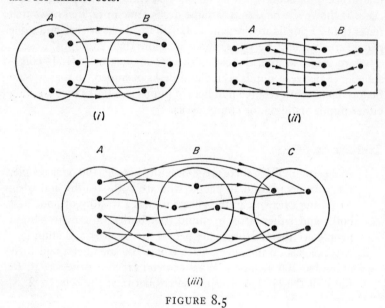

FIGURE 8.5

(i) for every set A, we have $A \sim A$
(ii) if $A \sim B$, then $B \sim A$
(iii) if $A \sim B$ and $B \sim C$, then $A \sim C$

To check (i), notice that the function Id_A which maps each element of A to itself is a mapping on A to A which is (1, 1) and onto. To check (iii), let $f: A \to B$, $g: B \to C$ be (1, 1) functions mapping A onto B and B onto C, respectively. Then the composition $g \circ f$ is (1, 1) onto C from A (see Sections 2.4 and 2.5 to refresh your memory on what this means). Finally, (ii) follows by considering the function

$h: B \to A$ inverse to the mapping $f: A \to B$. Notice that, in the definition, there is no claim of uniqueness for the function $f: A \to B$. If a set has more than one element in it, there are many ways of counting it!

For finite sets A, we have seen in Section 8.1 that there is a unique positive integer n such that:

$$A \sim \{1, 2, 3, \ldots, n\}.$$

Instead of saying that A has the same cardinal as the initial segment $\{1, 2, \ldots, n\}$ of N, we say briefly that A has cardinal n, or simply that there are n elements in A. The cardinal number n is thus the property of the class of all those sets which can be put into (1, 1) correspondence with the set $\{1, 2, \ldots, n\}$ (and with each other, see Fig. 8.6).

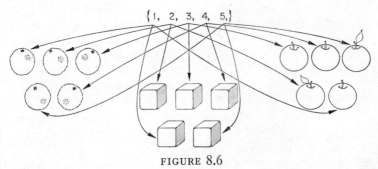

FIGURE 8.6

The paradox with which we opened this Chapter can now be understood in these terms, for if E is the set of even integers, D the set of odd integers, we proved that:

$$E \sim N \text{ and } D \sim N$$

so that E, D and N all have the same cardinal number. Mathematicians usually denote the cardinal of N by the symbol $\aleph_0$ (read aleph nought). Any set A which has cardinal $\aleph_0$ is said to be *denumerable*, for it is possible to arrange the elements of A as a sequence in which each element occurs once and only once. Let $f: N \to A$ be a (1, 1) function defined on N onto A, and put $f(n) = a_n \in A$. Then:

$$a_1, a_2, \ldots, a_n, \ldots$$

is such a sequence.

$\aleph_0$ is not just a very large number like 100^{100} or even 'infinity'. In fact we cannot use the word infinity to designate a number because it is too vague. We will see that from the point of view of counting, we can distinguish more than one kind of infinite number (in fact there are infinitely many different infinities!). Because the three distinct sets N, E and D can be linked by mappings which are $(1, 1)$ and onto we say that they have the same cardinal number, and that this number is called $\aleph_0$. Just as you teach a child the meaning of the number 'five' by showing him various sets with 5 members, the only way to think of the number $\aleph_0$ is that it is a common property of any denumerable set—that is, any set with $\aleph_0$ members in it.

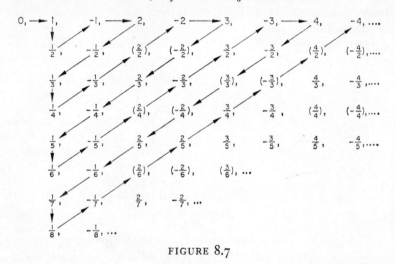

FIGURE 8.7

We have seen that proper subsets of N can have cardinal $\aleph_0$. It is fair to ask whether there are any sets larger than N which have the same cardinal. We shouldn't be easily surprised any more, but at first sight it is surprising that the set Q of all rational numbers has cardinal $\aleph_0$. On the number line, Q seems to have a lot more points in it than N, yet we can prove $Q \sim N$ by actually describing a $(1, 1)$ correspondence between these sets. There are many ways of doing this. Let us arrange the rational numbers as shown in Fig. 8.7, taking all the integers, positive and negative, in the first row, then all fractions with denominator 2 in the second row, and so on. We put brackets round those fractions for which the numerator and denominator have a common factor. If we delete these from our array of fractions, then each rational number will occur once and only once in the array. All we need do now is to give a method of rearranging this array as a single sequence in which every element occurs once and only once. The arrows in Fig. 8.7 indicate one way of achieving this to give the sequence:

$0, 1, \frac{1}{2}, -1, 2, -\frac{1}{2}, \frac{1}{3}, \frac{1}{4}, -\frac{1}{3}, -2, 3, \frac{2}{3}, -\frac{1}{4}, \frac{1}{5}, \frac{1}{6}, -\frac{1}{5}, -\frac{2}{3}, \frac{3}{2}, -3, 4, -\frac{3}{2}, \ldots$

The (1, 1) correspondence is not a simple one, but it is nevertheless defined since (a) we can find the nth term in this sequence and (b) each rational number p/q will be somewhere in the array of Fig. 8.7 and, after a finite number of steps, it will be reached as an element of the sequence.

You may begin to suspect that all sets which are not finite have the same cardinal number. The proof that this is false is again due to Cantor: it consists of proving that nonsense results if we assume that all infinite sets have cardinal $\aleph_0$. Consider the set I consisting of all real numbers x such that $0 < x < 1$. We will show that the cardinal of I is not $\aleph_0$ by showing that it is impossible to find a sequence $\{x_n\}$ of points of I in which every real number between 0 and I occurs. We saw in Chapter 4 that there is a unique representation of each $x \in I$ as an infinite decimal:

$$x = 0 \cdot a_1 a_2 a_3 \ldots a_n \ldots$$

not ending in 9 recurring. Suppose, if possible, that $\{x_n\}$ is a sequence containing every point of I. Express each number x_n as an infinite decimal so that:

$$x_1 = 0 \cdot a_{11} a_{12} a_{13} \ldots a_{1n} \ldots$$
$$x_2 = 0 \cdot a_{21} a_{22} a_{23} \ldots a_{2n} \ldots$$
$$x_3 = 0 \cdot a_{31} a_{32} a_{33} \ldots a_{3n} \ldots$$
.
.
.
$$x_k = 0 \cdot a_{k1} a_{k2} a_{k3} \ldots a_{kn} \ldots$$
.
.
.

each pair k, n of positive integers a_{kn} is the nth digit in the decimal expansion of the number x_k. We form a real number z such that

$0 < z < 1$ and z is different from all the x_k, by defining the digits in the decimal expansion of z as follows:

If a_{kk} is one of 0, 1, 2, 3, 4, put $b_k = 8$
if a_{kk} is one of 5, 6, 7, 8, 9, put $b_k = 1$

This determines the kth digit for each integer k, so that we have defined:

$$z = 0 \cdot b_1 b_2 \ldots b_k \ldots$$

Now, for each integer n, $z \ne x_n$ since the nth place in the expansion of z is b_n, which is certainly different from a_{nn}, which is the nth digit in the expansion of x_n. This contradiction proves that the cardinal of I is not $\aleph_0$.

It is usual to say that I has the cardinal c of the *continuum*. There is a sense in which $\aleph_1$ is the smallest infinite cardinal. So far we have shown that c is different from $\aleph_0$, so that it is a bigger cardinal number. The different notation c, $\aleph_0$ is explained only by the fact that we cannot tell whether c is the 'next' cardinal bigger than $\aleph_0$—and so there is no good reason for calling it $\aleph_1$. All we do know is that we can recognize many sets as having $\aleph_0$ elements and many others as having c elements and we have proved that $c \ne \aleph_0$.

Example 2. Just as there are many very different sets with cardinal $\aleph_0$, there are also many different sets with cardinal c. First, for any pair $a < b$ of real numbers, it is easy to see that:

$$(a, b) = \{x \in R \mid a < x < b\}$$

has the same cardinal as $(0, 1)$. We illustrate the $(1, 1)$ correspondence in the case where $b - a > 1$ in Fig. 8.8. Clearly each point x in the upper interval $(0, 1)$ corresponds to a unique point y in the lower interval (a, b), and vice versa.

It is more surprising that $(0, 1) \sim R$ so that the cardinal number of the set of all real numbers is also c. We can illustrate a correspondence which proves this by Fig. 8.9. Here we think of $(0, 1)$ as a segment broken at its mid-point ($\frac{1}{2}$) which is placed at O on the number line R. The two pieces AO, OB of $(0, 1)$ are then bent to lie at $45°$ to the number line, as shown. P is a point midway between A and B, and the correspondence is obtained by 'projecting' from the point P through OA or OB. Each $x \in (0, 1)$ then projects into a unique point y on the number line and the function is $(1, 1)$ and onto.

CARDINAL NUMBER

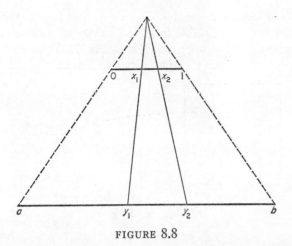

FIGURE 8.8

Even more surprising is that the cardinal number of the set of points in a square is also **c**. For we can think of the square S as consisting of those number pairs (x, y) with $0 < x < 1$, $0 < y < 1$, and we can exhibit a (1,1) correspondence between S and I, the interval $\{z \in R \mid 0 < z < 1\}$, as in Fig. 8.10. We define the function $f : I \to S$ by using the expansion of any real number as an infinite decimal. Suppose:

$$x = 0 \cdot a_1 a_2 \ldots a_n \ldots$$
$$y = 0 \cdot b_1 b_2 \ldots b_n \ldots$$

are infinite decimal expansions of x and y. Put:

$$z = 0 \cdot a_1 b_1 a_2 b_2 \ldots a_n b_n \ldots$$

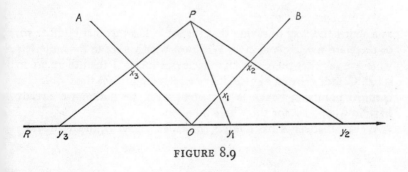

FIGURE 8.9

Then z is an infinite decimal in which the digits in odd places come from x, those in even places from y. Thus $0 < z < 1$, so we have set up a mapping $f: S \to (0, 1)$, which is clearly (1, 1) and onto, though it is not easy to picture the exact correspondence.

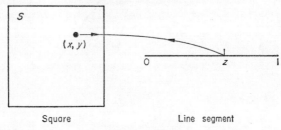

FIGURE 8.10

Exercise 8b

1. Show that the set A of those positive integers which are perfect squares is denumerable.
*2. Show that the set of points L in the plane of the form (x, y) with $x, y \in N$ is denumerable. (Hint: Use the same kind of diagonal argument which sufficed to show that the set of rationals is denumerable.)
*3. Show that the set of all points in the plane has cardinal **c**. (Hint: first set up a (1, 1) correspondence between the plane and the unit square S considered in Example 3.)

8.3 Comparing cardinals

Let us use:
$$\|A\| = \mathbf{m}$$
as a shorthand way of saying that the set A has cardinal number **m**, or there are **m** elements in A. We know what we mean for finite sets when we say that there are more elements in set A than there are in set B. Cantor showed how to define a relation 'greater than' or $>$ for cardinal numbers, which is consistent with the notion we already have in Chapter 3 for the whole numbers.

We say that set A has a *larger cardinal* than set B, in shorthand:
$$\|A\| > \|B\|$$

COMPARING CARDINALS

if there is a subset $A_1 \subset A$ such that $A_1 \sim B$, but there is no subset $B_1 \subset B$ with $A \sim B_1$.

For a pair of finite sets A, B with $\|A\| = 8$, $\|B\| = 5$, we can illustrate this ordering as in Fig. 8.11. It is clear that we cannot hope to set up a (1, 1) correspondence between the whole of A and a part of B, so $8 > 5$. This example makes the whole question of comparing cardinals seem trivial—so why do we bother with a complicated definition? The answer is that, for infinite sets, we need to be much more careful as the subsequent discussion will show.

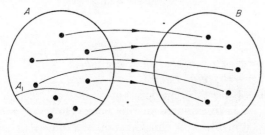

FIGURE 8.11

We have defined $>$ as a relation between pairs of cardinal numbers: let us check that it has the property of an order relation:

$$\mathbf{m} > \mathbf{n}, \mathbf{n} > \mathbf{p} \Rightarrow \mathbf{m} > \mathbf{p}$$

Suppose:

$$\mathbf{m} = \|A\| > \|B\| = \mathbf{n}, \text{ and } \|B\| > \|C\| = \mathbf{p}$$

Then choose $A_1 \subset A$ and $B_1 \subset B$ such that $A_1 \sim B$ and $B_1 \sim C$. Let $g : B \to A_1$ be the (1, 1) correspondence between A_1 and B and put $A_2 = \{x \in A_1 \mid x = g(y), y \in B_1\}$, i.e., A_2 is the image of B_1 under this correspondence. If $h : C \to B_1$ is a (1, 1) correspondence between C and B_1, then the composition $g \circ h : C \to A_2$ defines a (1, 1) correspondence between A_2 and C, so that $A_2 \subset A$, $A_2 \sim C$. To complete the argument we have to show that $A \sim C_1$ is impossible for any subset $C_1 \subset C$. Suppose $f : A \to C_1$ is a (1, 1) correspondence between A and C_1. Then $h \circ f : A \to B_2$ will be a (1, 1) correspondence between A and $B_2 = \{y \in B \mid y = h(z), z \in C_1\}$ the image in B of the set C_1 in C. This gives $A \sim B_2 \subset B$ which is impossible. Hence $\mathbf{m} > \mathbf{p}$.

Note that if m, n are finite positive integers then:

$$\mathbf{m} = \|\{1, 2, \ldots, m\}\|; \mathbf{n} = \|\{1, 2, \ldots, n\}\|$$

If $m > n$, in the sense of Chapter 3, then $\{1, 2, \ldots, m\} \supset \{1, 2, \ldots, n\}$; and there cannot be any subset of $\{1, 2, \ldots, n\}$ which has the same cardinal as $\{1, 2, \ldots, m\}$ for, if there were, we could take a box of n bricks, throw some of them away and end up with a box containing m bricks. This means that in the sense of our definition of ordering between cardinals we also have **m > n**. Our new definition is thus an extension of the old.

The definition makes sense because of the following result due to a mathematician called Bernstein.

THEOREM II

If A, B are any sets such that there are subsets $A_1 \subset A$, $B_1 \subset B$ with $A_1 \sim B$ and $A \sim B_1$, then A and B have the same cardinal, $A \sim B$.

The proof of this theorem does not require any techniques other than those we have explored in this book. However, we shall not include it, as it involves the consideration of a somewhat complicated function. The interested reader will find a proof in any advanced book on set theory, *e.g.*, see p. 22 of Kamke's, 'The Theory of Sets'. The theorem is extremely useful in determining the cardinal number of those sets which arise in practice.

Example 3. Suppose that A is any set and that $A_2 \subset A_1 \subset A$, but $\|A_2\| = \|A\|$; it follows then that $\|A_1\| = \|A\|$, so that all three sets have the same cardinal. We can deduce this from Bernstein's theorem for $A \sim A_2$ which is a subset of A_1, since A and A_2 have the same cardinal, and $A_1 \sim A_1$ which is a subset of A, using the identity mapping, so $A_1 \sim A$.

In particular, any set E of real numbers such that $E \supset$ some interval (a, b) must have cardinal **c**.

Example 4. A set E is infinite if, and only if there is a subset $F \subset E$ such that $\|F\| = \aleph_0$. For E has the property that, however large an integer n we are given, it is possible to pick a subset of E with n elements, and there will still be part of E left. By induction we can pick a sequence:

$$x_1, x_2, \ldots, x_n, \ldots$$

of distinct elements of E. Let F be the set of points in this sequence.

Example 5. A set E is infinite if and only if there is a proper subset $G \subset E$ such that $G \sim E$. So the paradox of Section 8.1 is a characteristic of all infinite sets, namely that any infinite set has the same number of elements as a part of itself.

Use Example 4 to find a subset $F \subset E$ which can be written as a sequence:
$$x_1, x_2, \ldots, x_n, \ldots$$
Let H be the sequence of even elements of F:
$$x_2, x_4, \ldots, x_{2n}, \ldots$$
Then we know that $F \sim H$. If we put:
$$B = (E - F) \cup H,$$
then B is a proper subset of E, since it does not contain any of the points $x_1, x_3, \ldots, x_{2n-1}, \ldots$ We can set up the (1, 1) correspondence on B to E by defining:
$$f(x) = x, \text{ for } x \in (E - F)$$
and $\qquad f(x_{2n}) = x_n$, for $x_{2n} \in H$, $n = 1, 2, \ldots$

Now that we know we can sometimes (we do not know whether it is always possible) compare two infinite cardinals, there are immediate problems.

Example 6. Is there any cardinal **m** such that $\mathbf{c} > \mathbf{m} > \aleph_0$?

It is clear from our definition of ordering that $\mathbf{c} > \aleph_0$: this question really asks are there any sets on the real line which are infinite, cannot be arranged as a sequence, and cannot be put in (1, 1) correspondence with R. The answer to this question is not known, and many mathematicians believe that it cannot be decided by logical argument from the other axioms of mathematics. The assumption that no such set exists is called the *continuum hypothesis*: it is known that this assumption does not lead to a contradiction with the other usual rules of logical reasoning.

Example 7. Are there larger cardinals than **c**?

Here the simple answer is 'yes'. In fact, for any set A, the set whose elements consist of all the possible subsets of A is called the *power set* of A. It can be proved that:
$$\|2^A\| > \|A\|$$
for every set A, where 2^A is the power set of A. The proof of this is

again rather complicated; the reader is referred to p. 21 of Kamke's, 'The Theory of Sets'.

This means that we can generate as many infinite cardinals as we please for $c > \aleph_0$, $f > c$ etc., where f is the cardinal number of the class of all subsets of the real number system.

Example 8. Are all cardinals comparable? In other words, given two sets A, B is it always true that either:
   (i) there is a subset $A_1 \subset A$ with $A_1 \sim B$ or
   (ii) there is a subset $B_1 \subset B$ with $A \sim B_1$?

We have seen that if both (i) and (ii) are true, then $A \sim B$, but we do not know that at least one of (i) or (ii) is true. This is not a question which can be answered in terms of cardinal numbers. It can be answered, however, if we introduce an additional axiom or assumption into our logical system. There are various equivalent forms of this assumption, but perhaps the easiest intuitively is the well ordering axiom, which states:

Given any set E it is possible to define an order relation on E such that:
   (i) *E is simply ordered; that is, given* $x, y \in E$ *either* $x = y$ *or* $x > y$ *or* $y > x$.
   (ii) *E is well ordered; that is, given any subset A of E either* $A = \emptyset$ *or A has a least member*.

Note that the set N of positive integers with the usual ordering is well ordered, but it is not possible to define constructively an order relation on R, the set of real numbers which makes R well ordered (the usual order in R is not a well ordering). If the well ordering axiom is true, then all cardinals are comparable.

Before we leave the question of cardinal numbers, it is worth asking whether the arithmetic operations of addition $(a + b)$ multiplication $(a.b)$ and raising to a power (a^b) can be carried out for cardinal numbers, since we know they can be defined in the set N of positive integers, *i.e.*, for finite cardinals. Cantor showed that each of these operations could be defined for all pairs of cardinals **a**, **b** in such a way that they give the familiar answer for integers. However, it is not possible to define subtraction or division for infinite cardinals in order to give a unique answer. For details the reader should consult chapter 2 of Kamke's, 'The Theory of Sets'.

Exercise 8c

1. Show that the set of points inside a circle of radius 2 has cardinal **c**.
2. Show that, if A, B are disjoint sets such that $\|A\|$ is finite, $\|B\| = \aleph_0$ then $\|A \cup B\| = \aleph_0$.
3. Show that, if A, B are disjoint sets such that $\|A\| = \aleph_0 = \|B\|$, then $\|A \cup B\| = \aleph_0$.
*4. Suppose we have a ball-bearing factory which is producing balls in order at an ever increasing rate. The balls are numbered $1, 2, 3, \ldots, n, \ldots$ and are released down a chute C into an infinite store B from which they are taken in the right order (see Fig. 8.12), as follows:

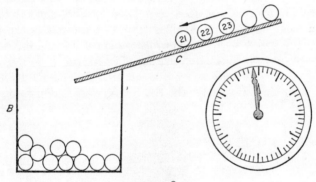

FIGURE 8.12

At 1 min to midnight, balls numbered 1 to 10 enter B, ball No. 1 is removed

At $\frac{1}{2}$ min to midnight, balls 11–20 enter B, ball 2 is removed

At $\frac{1}{3}$ min to midnight, balls 21–30 enter B, ball 3 is removed

At $\frac{1}{n}$ min to midnight, balls $[10(n-1)+1]$ to $10 \cdot n$ enter B, and ball n is removed.

How many balls are in the store B at midnight?

Example 9. The set of irrational real numbers is not denumerable. For $R = Q + S$ where Q is the set of rationals, and S is the set of irrational numbers. We know that $\|Q\| = \aleph_0$ so if $\|S\| = \aleph_0$ then by question 3 in Exercise 8c, we would have $\|R\| = \aleph_0$, which is false.

9 | OVER THE TOP: GOING TO THE LIMIT

9.1 Some paradoxes

In the fifth century BC, Zeno of Elea described several paradoxes. The first of these which we consider is called 'Achilles and the tortoise' and can be described as follows. Achilles and the tortoise have a race, and because Achilles can run ten times as fast as the tortoise it is agreed to give the tortoise a start of 100 metres. Suppose for the purpose of calculation that Achilles can run at 10 metres per second, and the tortoise at 1 metre per second. This means it takes Achilles 10 seconds to cover the 100 metres to the point from which

FIGURE 9.1

the tortoise started (Fig. 9.1). But in this time the tortoise has gone 10 metres ahead, so Achilles must keep going. He takes 1 second to cover this distance only to find the tortoise has gone another metre. Achilles takes $\frac{1}{10}$ of a second to get to this point, but again finds the tortoise has moved on $\frac{1}{10}$ of a metre. The gap is narrowing, but it looks as though Achilles will never catch up with the tortoise, because every time he runs to where the tortoise was, he discovers that the tortoise has moved on a bit further!

Another of Zeno's paradoxes indicates that motion is impossible. Suppose we wish to move from a point P to a point Q along a straight line (see Fig. 9.2). We must first go half the distance from P to Q—

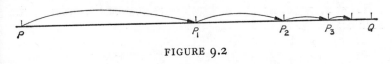

FIGURE 9.2

say to the mid-point, P_1; then half the remaining distance, then half the distance then remaining, then half the distance then remaining, and so on. In spite of the fact that successive distances are getting smaller, each definitely takes a positive time to cover. This means that we can never get to Q, as we would require an infinite number of these intervals of time. Zeno argued that the sum of an infinite number of intervals must be infinite. This argument can be repeated to show that you would never get from P to P_1 or in fact from P to any point on the way to Q—hence motion is impossible!

Another paradox results from consideration of a bouncing ball. If we drop a steel ball bearing, or a golf ball on a hard, horizontal surface it will bounce back up to a slightly lower height than it started from. If we leave it to move freely it will fall and bounce again. Each time it hits the ground it immediately leaves it and takes some time to get back again. It cannot come to rest after a finite number, n, of bounces because the 'last' time it comes down it is moving and so will bounce up again. Since each bounce takes a positive amount of time, and there are an infinite number of bounces, the ball will therefore never come to rest.

Your intuition may by now be good enough to see the answer to each of these apparent fallacies. In fact, the key to the problem is the possibility of defining addition for an infinite collection of numbers. We shall indicate how this can sometimes be done in Section 9.3, but let us first consider paradoxes which can result from infinite addition sums—if we are not careful. This will show us that we need to think very carefully if we are going to justify the extension of the notion of addition to an infinite number of terms.

F

Example 1. Consider this series:

$$S = 1 - 2 + 4 - 8 + 16 - 32 + 64 - 128 + \ldots$$
$$= 1 - 2(1 - 2 + 4 - 8 + 16 - 32 + 64 - \ldots)$$
$$= 1 - 2S$$

So that:

$$3S = 1 \text{ or } S = \tfrac{1}{3}$$

However, we could put in brackets to change the order of the addition:

$$S = 1 + (-2 + 4) + (-8 + 16) + (-32 + 64) + \ldots$$
$$= 1 + 2 + 8 + 32 + \ldots$$

so that S is bigger than any number we care to name. Or:

$$S = (1 - 2) + (4 - 8) + (16 - 32) + (64 - 128) + \ldots$$
$$= -1 - 4 - 16 - 64 - \ldots$$

so that S is less than any large negative number we care to name. All we have done is to insert brackets in the above—indicating the order in which the operations are to be carried out. We have only used operations which we know are completely justified when we are adding (and subtracting) a finite number of terms.

You may think that the cause of the trouble in Example 1 is that the terms being added are getting larger. However, trouble can arise even if it the terms get small.

Example 2. Put:

$$L = 1 - \tfrac{1}{2} + \tfrac{1}{3} - \tfrac{1}{4} + \tfrac{1}{5} - \tfrac{1}{6} + \tfrac{1}{7} - \tfrac{1}{8} + \ldots$$

Multiply by 2, giving:

$$2L = 2 - 1 + \tfrac{2}{3} - \tfrac{1}{2} + \tfrac{2}{5} - \tfrac{1}{3} + \tfrac{2}{7} - \tfrac{1}{4} + \ldots$$

Now group the terms with the same denominator:

$$2L = (2 - 1) - \tfrac{1}{2} + (\tfrac{2}{3} - \tfrac{1}{3}) - \tfrac{1}{4} + (\tfrac{2}{5} - \tfrac{1}{5}) - \tfrac{1}{6} + \ldots$$
$$= 1 - \tfrac{1}{2} + \tfrac{1}{3} - \tfrac{1}{4} + \tfrac{1}{5} - \tfrac{1}{6} + \ldots$$
$$= L$$

We cannot have $2L = L$ unless $L = 0$ which it clearly is not since

$$L = (1 - \tfrac{1}{2}) + (\tfrac{1}{3} - \tfrac{1}{4}) + \ldots,$$

a sum of definitely positive numbers. (It can, in fact, be shown that $L = \log_e 2 = 0\cdot 69314\ldots$)

9.2 Beating the enemy

The sequence whose nth term $a_n = 1/n$:

$$1, \frac{1}{2}, \frac{1}{3}, \frac{1}{4}, \ldots, \frac{1}{n}, \ldots$$

has an intuitive property which we can describe as '$1/n$ approaches 0 as n grows larger'.

Let us try to analyse exactly what we mean by this. As we go further along the sequence the terms get smaller and smaller. After the 50th term they are all smaller than $\frac{1}{50}$; after the 1000th term they are all smaller than $\frac{1}{1000}$. None of the terms is actually equal to 0 (for infinity does not correspond to a term in the series because it is not an integer). However, if we go out far enough along the sequence all the terms will differ from 0 by as little as we please.

To make this precise, think of a game or a war with an adversary A.

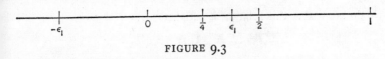

FIGURE 9.3

The rules of the war are as follows:

(i) A may choose a positive real number—it doesn't matter how small; call it $\epsilon_1 > 0$ (see Fig. 9.3).

(ii) In order to win the battle for ϵ_1 you have to find an integer N_1 large enough to ensure that the terms of the sequence are all within ϵ_1 of zero. In this case we can take for N_1 any integer bigger than $1/\epsilon$, since then for integers n:

$$n \geqslant N_1 \Rightarrow \frac{1}{n} \leqslant \frac{1}{N_1} < \epsilon_1.$$

(iii) The enemy may then start another battle by changing his ϵ_1 to a different number ϵ_2—so long as $\epsilon_2 > 0$.

(iv) You have to beat him on this occasion by choosing another integer N_2 large enough to make the terms of the sequence within ϵ_2 of zero.

(v) In order to win the war you have to win every battle: this means that no matter what positive number ϵ the adversary chooses, you have to be able to beat it by finding an integer N such that all the terms from N onwards are within ϵ of zero.

This notion of always being able to 'beat the enemy' is precisely what we mean by *convergence*, or *tending to a limit* for a sequence of real numbers. It applies to any sequence:

$$a_1, a_2, a_3, \ldots, a_n, \ldots$$

of numbers. We say that the sequence converges to the real number a as n tends to infinity, and write:

$$a_n \to \infty \text{ as } n \to \infty, \text{ or } \lim_{n \to \infty} a_n = a$$

if no matter what positive ϵ is chosen we can find an integer N such that, whenever $n \geqslant N$, a_n is within ϵ of the number a. This means that, though we do not know whether the terms:

$$a_1, a_2, \ldots, a_{N-1}$$

satisfy:

$$a - \epsilon < a_i < a + \epsilon$$

we do know that this relation is satisfied for:

$$i = N, N+1, N+2, \ldots$$

Another way of picturing convergence is to look at the graph of the sequence. This is the set of pairs (see Section 2.3):

$$(a_n, n) \qquad n = 1, 2, \ldots$$

In the war we have described when the enemy picks his $\epsilon < 0$, we can draw a pair of parallel lines at a distance ϵ on each side of the height a. In order to 'beat' ϵ we have to get the points of the graph in this strip (see Fig. 9.4). This means we have to find an N large enough to ensure that to the right of N there are no points of the graph outside this strip. If we can always do this, then:

$$a_n \to a \text{ as } n \to \infty.$$

The precise formulation we have given for the idea of convergence is quite subtle, and you should not be disappointed if you find it difficult to grasp fully. One needs quite a lot of experience of working with this notion before one feels at home with it.

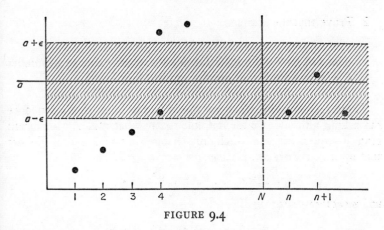

FIGURE 9.4

Example 3. The sequence given by:

$$a_n = \frac{n}{n+1}$$

converges to the limit 1 as n tends to infinity. For:

$$a_n = 1 - \frac{1}{n+1}$$

If the enemy chooses $\epsilon > 0$, we can take N to be the first integer larger than $\frac{1}{\epsilon}$. Then we shall have $1/N < \epsilon$, so that for any integer, n:

$$n \geqslant N \Rightarrow 0 < \frac{1}{n+1} < \epsilon$$

and $$1 - \epsilon < a_n < 1.$$

This gives a strategy which will win every battle, which means $a_n \to 1$ as $n \to \infty$.

Exercise 9a

1. Prove that the sequence:

$$a_n = \frac{n^2 + n}{n^3 + n}$$

has limit 0. (Hint: a_n lies between 0 and $2/n$.)

2. Prove that the sequence:
$$a_n = (-1)^n$$
does not have a limit. (Hint: let the enemy choose a number $\epsilon = \tfrac{1}{2}$, and show that no strategy will beat him.)

There are occasions when we would like to know that a sequence has a limit although we are not able to say what it is. An important kind of sequence is one which only changes in one direction. We say that $a_1, a_2, \ldots, a_n, \ldots$ is *monotone increasing* if:

$$a_1 \leq a_2 \leq a_3 \leq \ldots \leq a_n \leq a_{n+1} \leq \ldots$$

and *monotone decreasing* if:

$$a_1 \geq a_2 \geq a_3 \geq \ldots \geq a_n \geq a_{n+1} \geq \ldots$$

Such sequences can approach a limit from one side only. Only two kinds of behaviour are possible for a monotone sequence. It may run away completely like

$$1, 4, 9, 16, 25, \ldots$$

where $a_n = n^2$, or:

$$2, 3, 5, 7, 11, 13, \ldots$$

where a_n is the nth prime number. If this does not happen, then the sequence is bounded—that is, there is some real number K such that every term is smaller than K. It is often possible to find a bound K such that $a_n \leq K$ for all integers n, without being able to find the smallest such bound K. Now if you picture the sequence on the number line (see Fig. 9.5), it is moving to the right as n increases, but

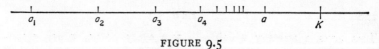

FIGURE 9.5

cannot get beyond K. Intuitively it must 'pile up' somewhere, and this will be the limit a of the sequence. To prove this requires a deep property of the real numbers which we discussed in Section 4.5.

Example 4. Consider the sequence defined by:

$$a_n = \left(1 + \frac{1}{n}\right)^n$$

BEATING THE ENEMY

If we expand this by the binomial theorem (Example 6 in Chapter 3, p. 48), we get $(n+1)$ terms as follows:

$$a_n = 1^n + n \cdot 1^{n-1}\left(\frac{1}{n}\right) + \frac{n(n-1)}{2} 1^{n-2}\left(\frac{1}{n}\right)^2 + \ldots + \left(\frac{1}{n}\right)^n$$

$$= 1 + 1 + \frac{1}{2}\left(1 - \frac{1}{n}\right) + \frac{1}{3!}\left(1 - \frac{1}{n}\right)\left(1 - \frac{2}{n}\right) + \ldots$$

$$+ \frac{1}{n!}\left(1 - \frac{1}{n}\right) \ldots \left(1 - \frac{n-1}{n}\right)$$

Here the $(k+1)$th term to be added is:

$$\frac{1}{k!}\left(1 - \frac{1}{n}\right)\left(1 - \frac{2}{n}\right) \ldots \left(1 - \frac{k-1}{n}\right)$$

which is $< 1/k!$, since all the terms in the brackets are smaller than 1. Now, since $k!$ is a product of the $(k-1)$ integers, $2.3.4 \ldots k$, each of which is at least 2, we have $k! \geqslant 2^k$ for all integers k, so that:

$$a_n \leqslant 1 + 1 + \tfrac{1}{2} + (\tfrac{1}{2})^2 + \ldots + (\tfrac{1}{2})^n$$
$$= 3 - (\tfrac{1}{2})^n < 3.$$

Hence for this particular sequence $a_1, a_2, \ldots a_n \ldots$ we have shown that $a_n < 3$ for all integers n.

The argument could clearly be improved to give a smaller value of K, but it is not obvious what is the smallest value which will work.

The set of real numbers in the sequence is bounded above (by K) so it has a least upper bound. It is not too hard to prove that this least upper bound, l, is the limit of the sequence.

A sequence which ultimately gets larger than any number K is said to tend to plus infinity as n tends to infinity. A monotone increasing sequence either tends to plus infinity or it converges to a limit.

Exercise 9b

If $r > 0$ and $a > 0$, we can add together n numbers to give:

$$S_n = a + ar + ar^2 + \ldots + ar^{n-1}.$$

This is called a *geometric progression* with n terms. Then, calculation shows that if $r \neq 1$:

$$S_n = a\frac{1 - r^n}{1 - r}$$

Thus the number S_n defines a sequence for $n = 1, 2, \ldots$ which is monotone increasing. Show that:

(a) if $0 < r < 1$, S_n converges to a limit;
(b) if $r \geqslant 1$, $S_n \to +\infty$, as $n \to \infty$.

Example 5. (see p. 52). If we count up the number of primes which are $< n$ and call this $f(n)$, we can define a sequence:

$$r_n = \frac{f(n)}{n} \log_e n$$

where $\log_e n$ stands for the Napierian logarithm (to the base e) of n. The celebrated prime number theorem, discovered, but not proved by Gauss, tells us that the sequence $r_1, r_2, \ldots, r_n \ldots$ converges to 1 as $n \to \infty$. There is a discussion of this on pp. 27–30 of Courant and Robbins, 'What is Mathematics?'

9.3 Accountant's nightmare

Have you ever added up a long column of figures and heaved a sigh of relief when you came to the end? What would you do if the column went on and on, never coming to an end? If one worked on such an infinite addition sum, the partial answers would keep getting larger as we add more terms. It might be obvious that the answer was going to keep on growing indefinitely—that is, that we could make it as large as we please by going on long enough: in this case we can say that our sum to infinity is infinite. The other alternative is that, after a while, we might develop a hunch that there is some number K such that the result of adding together any finite number of terms will never exceed this number. If this hunch is correct, then there ought to be an answer to our infinite addition sum. We won't be able to find it by adding, but we can get an approximate answer—to any prescribed degree of approximation—by adding a large finite number of terms.

There is a story about the inventor of chess who was offered a reward for his ingenuity by the Emperor of the day. The inventor asked to be given a grain of rice for the first square on a chessboard, two grains for the second square, four grains for the third, eight grains for the fourth, and so on. This seemed trivial to the Emperor, who ordered that he be given the reward. But a chessboard has 64 squares, so that the number of grains of rice in the reward was:

$$1 + 2 + 2^2 + \ldots + 2^{63} = 2^{64} - 1.$$

This is a very large number—expressed in figures it is:

$$18\ 446\ 744\ 073\ 709\ 551\ 615$$

—in fact the inventor was asking for more rice than ever existed on the Earth!

In this case it is quite obvious that the infinite sum:

$$1 + 2 + 2^2 + \ldots + 2^{n-1} + \ldots$$

must be infinite, for we keep adding larger and larger terms. Before there is to be any hope of a finite answer to our infinite addition sum we must have terms which are getting small. In fact in the notation of Section 9.2, if the terms $a_n > 0$, the infinite sum:

$$a_1 + a_2 + a_3 + \ldots + a_n + \ldots$$

cannot possibly be finite unless $a_n \to 0$ as $n \to \infty$. For, if this is false, the enemy must be able to find some number $\epsilon > 0$ which 'beats' us—that is, there are infinitely many of the terms $a_n > \epsilon$. This implies that ultimately, after k such terms have been counted, the sum will be bigger than $k\epsilon$, so that it can be made as large as we wish.

However, the condition $a_n \to 0$ is not sufficient to ensure that we can find an infinite sum for $a_1 + a_2 + \ldots + a_n + \ldots$ For example, let $a_n = 1/n \to 0$ as $n \to \infty$, as we saw in Section 9.2. That is we are considering the infinite addition sum:

$$1 + \tfrac{1}{2} + \tfrac{1}{3} + \tfrac{1}{4} + \tfrac{1}{5} + \tfrac{1}{6} + \tfrac{1}{7} + \tfrac{1}{8} + \ldots$$

Here the terms we are adding get small quite quickly—can we add them together? Let us group the terms in brackets as follows:

$$1 + (\tfrac{1}{2} + \tfrac{1}{3}) + (\tfrac{1}{4} + \tfrac{1}{5} + \tfrac{1}{6} + \tfrac{1}{7}) + (\tfrac{1}{8} + \tfrac{1}{9} + \ldots + \tfrac{1}{15}) + \ldots$$

so that the first bracket contains two terms, the second four terms and the nth bracket contains 2^n terms starting with 2^{-n}. The smallest of the terms in the nth bracket is the last, namely:

$$\frac{1}{2^{n+1} - 1}$$

which is larger than $1/2^{n+1}$ ($= 2^{-n-1}$). This means that all the terms in the nth bracket are larger than 2^{-n-1}, and each of the brackets has a sum bigger than $2^n . 2^{-n-1} = \frac{1}{2}$.

By continuing far enough to include all the terms up to the nth bracket, this will give an answer bigger than:

$$1 + \underbrace{\tfrac{1}{2} + \tfrac{1}{2} + \ldots + \tfrac{1}{2}}_{n \text{ terms}} = 1 + \tfrac{1}{2}n$$

which will ultimately grow bigger than any prescribed number K. Thus the infinite sum:

$$1 + \frac{1}{2} + \frac{1}{3} + \frac{1}{4} + \frac{1}{5} + \ldots + \frac{1}{n} + \ldots$$

gives an infinite answer. This is called the *harmonic series*. Although the infinite addition is impossible in this case, the sum grows extremely slowly. In fact, suppose you were challenged by the 'enemy' to make the sum bigger than 10. Since the brackets certainly add up to less than 1 you would have to go on at least to the end of the ninth bracket, which ends with $1/1023$—that is, you would have to add more than 1000 terms. To get a sum bigger than 100, you would have to add more than 10^{30} terms! Even with a modern high-speed computer, this would not be a practical proposition. Yet it is logically possible to add together 10^{30} terms, or any finite number of terms, so we conclude that the harmonic series must have an infinite sum.

We can understand both of Zeno's paradoxes in the context of an infinite sum which works out. Take Achilles and the tortoise as an example. Here the time taken by Achilles to cover the first 100 metres is 10 seconds. In order to get to the point the tortoise has now reached, he has to run for just 1 second. The next stage takes $\frac{1}{10}$ second, then $\frac{1}{100}$ second, and so on. That is the total time needed to catch up with the tortoise is, in seconds:

$$10 + 1 + \frac{1}{10} + \frac{1}{10^2} + \frac{1}{10^3} + \ldots$$

Now adding the first n of these terms gives:

$$10 \cdot \frac{1-\left(\frac{1}{10}\right)^n}{1-\left(\frac{1}{10}\right)} = \frac{100}{9} - \frac{100}{9}\left(\frac{1}{10}\right)^n$$

This means that, however many terms we add, we shall always get an answer less than $11\frac{1}{9}$, and we can get as close as we like to this number by taking sufficiently many terms. We say that the *sum* of our infinite series:

$$10 + 1 + \frac{1}{10} + \frac{1}{10^2} + \ldots$$

is $11\frac{1}{9}$. Thus it takes Achilles precisely $11\frac{1}{9}$ seconds to catch up with the tortoise—a result which agrees with our intuition about the situation for in 1 second, Achilles catches up 9 metres and they start 100 metres apart so it ought to take $100/9$ seconds to catch up.

Consider the paradox of the bouncing ball. If the ball is dropped from a height of h_1 metres and we ignore air resistance, Newton's laws tell us that it will hit the flat surface with a velocity of v_1 metres per second, where:

$$v_1^2 = 2gh_1$$

and g is the rate of acceleration due to gravity, *i.e.*, the rate at which a body falling to earth accelerates: if velocity is measured in metres per second, then g is approximately 9·8 metres per second per second. Experiment shows that there is a constant $k < 1$ (depending on the ball and the surface) such that the ball leaves the surface with a velocity v_2 metres per second, where:

$$v_2 = kv_1.$$

If we ignore air resistance it will come back to hit the surface with the same velocity it had when it left, namely v_2, and will take off again with a velocity v_3, where:

$$v_3 = kv_2 = k^2v_1.$$

The time taken between the first and second occasions the ball is in contact with the surface can be calculated from Newton's laws. It is:

$$t_2 = \frac{2}{g}v_2 = \frac{2}{g}k_1v.$$

Repeating the argument shows that the time between the $(n-1)$th and nth occasion of contact is:

$$t_n = \frac{2}{g} k^{n-1} v_1.$$

If t_1 is the time to the first contact, the total time before rest is:

$$t_1 + t_2 + t_3 + \ldots + t_n + \ldots$$
$$= \frac{1}{g} v_1 [1 + 2k + 2k^2 + \ldots + 2k^{n-1} + \ldots]$$

The sum of the first n terms of this series is:

$$\frac{2}{g} v_1 \left(\frac{1-k^n}{1-k} - \tfrac{1}{2} \right)$$

which is always smaller than:

$$T = \frac{2}{g} v_1 \left(\frac{1}{1-k} - \tfrac{1}{2} \right)$$
$$= \frac{1}{g} \frac{k+1}{1-k} v_1$$

and gets closer to this number T as n becomes large, since k^n gets very small. This number T is the total time (in seconds) it takes for the ball to come to rest.

Exercise 9c

Resolve Zeno's second paradox—that motion is impossible—by a similar argument.

9.4 Summing an infinite series

We can now give a precise definition for the operation fo finding the sum of:

$$a_1 + a_2 + a_3 + \ldots + a_n + \ldots$$

when this exists—where the terms $a_1, a_2, \ldots$ are all real numbers, positive or negative or zero. First form the sequence:

$$S_n = a_1 + a_2 + a_3 + \ldots + a_n$$

SUMMING AN INFINITE SERIES

for each positive integer n by taking the ordinary sum of the first n terms of the series. We call S_n the partial sum of the series. If this sequence S_n converges to a limit S as $n \to \infty$, then we say that S is the sum of the series.

Recall what this means in terms of beating the enemy. For any $\epsilon > 0$, however small, we can find an integer N so that the approximation S_n obtained by adding the first n terms is always within ϵ of S whenever $n \geqslant N$.

Note that you always obtain S_{n+1} from S_n by adding the term a_{n+1}. Hence if all the terms of the series are positive we have $S_{n+1} \geqslant S_n$ for all n, which means that the sequence S_n is monotone increasing. Its behaviour as n increases must therefore be simple:

(i) either S_n grows larger than any given number—we say that the series *diverges* to plus infinity in this case

(ii) or S_n is bounded by some number K, in which case it converges to a limit S which must be the sum of the series.

Example 6. A very important series in mathematics was first discussed by Euler:

$$1 + \frac{1}{1!} + \frac{1}{2!} + \frac{1}{3!} + \ldots + \frac{1}{n!} + \ldots$$

where $n!$ is the product of the first n integers. Note that:

$$\frac{1}{n!} = \frac{1}{2}.\frac{1}{3}.\frac{1}{4} \cdots \frac{1}{n} < \underbrace{\frac{1}{2}.\frac{1}{2} \cdots \frac{1}{2}}_{(n-1) \text{ factors}} = \frac{1}{2^{n-1}}$$

so that:

$$S_n = 1 + \frac{1}{1!} + \frac{1}{2!} + \frac{1}{3!} + \ldots \frac{1}{n!}$$

Replacing each term $\frac{1}{k!}$ by the larger term $\frac{1}{2^{n-1}}$ gives:

$$S_n < 1 + 1 + \frac{1}{2} + \frac{1}{2^2} + \ldots \frac{1}{2^{n-1}}$$

$$= 3 - \frac{1}{2^{n-1}} < 3$$

using the calculation of Example 4. This means that $S_n < K = 3$ for all n; and it is a monotone increasing sequence. It must converge to a limit, which we call e:

$$S_n \to e \text{ as } n \to \infty.$$

Written as a decimal $e = 2\cdot 718 \ldots$ It is easy to show that this real number is not rational; that is, you cannot find integers p, q such that $e = \frac{p}{q}$. In fact it can be shown that there is no polynomial equation:

$$a_n x^n + a_{n-1} x^{n-1} + a_{n-2} x^{n-2} + \ldots a_1 x + a_0 = 0$$

with the a_i all integers for which e is a solution. This means e is not of the form $\sqrt{(p/q)}$ or $\sqrt[n]{k}$.

Example 7. Consider the series obtained from the harmonic series by making the terms alternatively positive and negative:

$$1 - \frac{1}{2} + \frac{1}{3} - \frac{1}{4} + \frac{1}{5} - \frac{1}{6} + \ldots + \frac{(-1)^{n-1}}{n} + \ldots$$

If we take the sum of an even number of terms:

$$S_{2k} = \left(1 - \frac{1}{2}\right) + \left(\frac{1}{3} - \frac{1}{4}\right) + \ldots + \left(\frac{1}{2k-1} - \frac{1}{2k}\right)$$

by putting the terms in pairs in brackets we see that S_{2k} is the sum of k brackets each of which is positive. Further, changing k to $(k+1)$ has the effect of adding one more positive bracket so that $S_{2k+2} > S_{2k}$. Similarly the sum of an odd number of terms:

$$S_{2k+1} = 1 - \left(\frac{1}{2} - \frac{1}{3}\right) - \left(\frac{1}{4} - \frac{1}{5}\right) - \ldots - \left(\frac{1}{2k} - \frac{1}{2k+1}\right)$$

is obtained by subtracting k brackets each of which is positive. Now if we change k to $(k+1)$ we must take away one more positive

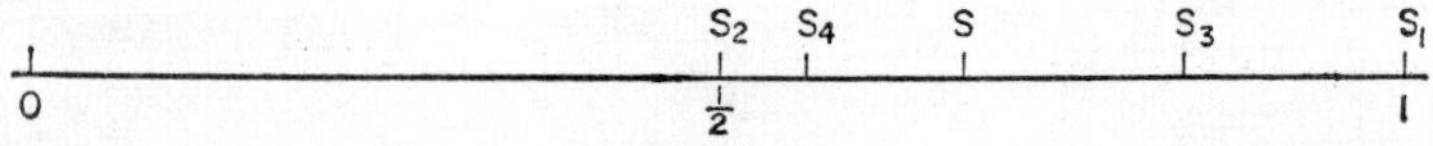

FIGURE 9.6

bracket, so $S_{2k+3} < S_{2k+1}$. Calculate the first few partial sums, and mark them on a number line (see Fig. 9.6). The sequence:

$$S_2, S_4, \ldots, S_{2k}, \ldots$$

is monotone increasing and bounded above by 1 for:

$$S_{2m} = 1 - \left(\frac{1}{2} - \frac{1}{3}\right) - \left(\frac{1}{4} - \frac{1}{5}\right) - \ldots - \left(\frac{1}{2k-2} - \frac{1}{2k-1}\right) - \frac{1}{2m}$$

It must therefore converge to some limit S. But the difference between S_{2k} and S_{2k+1} is only $1/(2k+1)$, which can be made as small as we please. Hence:

$$S_1, S_3, S_5, \ldots, S_{2k+1}, \ldots$$

also converges to the same limit S, which by definition is the sum of the series. It is perhaps of interest to note that this number:

$$S = \log_e 2 = 0 \cdot 693147 \ldots$$

The argument we have used in Example 7 will apply to any series in which the terms decrease in magnitude towards the limit zero and alternate in sign. In Example 2 we manipulated this series for $\log_e 2$ by multiplying every term by 2 (which is justified) and then rearranging the terms in a different order (which cannot be justified). By doing this we obtained nonsense, so it is clear that, in general, the order of the terms in an infinite series can affect the answer.

The question as to whether the order matters depends on whether the series

$$|a_1| + |a_2| + |a_3| + \ldots + |a_n| + \ldots,$$

obtained by replacing each term by a positive term of the same absolute size, converges or not. It is not too hard, but beyond the scope of this book, to prove that if the series of absolute values converges and the original series converges to a sum S, then all possible rearrangements will converge to the same sum S. Now in the case of Example 7 we know that the series of absolute values does not converge, so it should not surprise us that rearrangements give a different sum. In fact Riemann proved in 1854 the remarkable theorem that given a series:

$$a_1 + a_2 + a_3 + \ldots + a_n + \ldots$$

which converges to a sum S, but is such that:

$$|a_1| + |a_2| + |a_3| + \ldots + |a_n| + \ldots$$

diverges to $+\infty$, it is possible to rearrange the order of the terms in the series to make it converge to any specified number or to make it diverge to $+\infty$, or to $-\infty$. This explains the paradox of Example 2.

For a deeper discussion of the theory of convergence, both for sequences and series, consult Chapters 2 and 5 of Burkill's, 'A First Course in Mathematical Analysis'.

Exercise 9d

Show that the series:

$$1 + \frac{1}{2^2} + \frac{1}{3^2} + \frac{1}{4^2} + \ldots + \frac{1}{n^2} + \ldots$$

converges to a finite sum. (Hint: group the terms from $n = 2^k$ to $n = 2^{k+1} - 1$ in a bracket and estimate its size.)

9.5 How to measure the size of a set

Suppose we have a space S and we want to measure the exact 'size' of subsets $A, B, C, \ldots$ of S. Intuitively we could think of the sets as being composed of matter, so that they could be weighed exactly. If the sets were a nice shape we might try to calculate their area, or volume. Without being specific, let $f(A)$ denote some measurement of A, giving a real number. What sort of properties would we expect of this measuring function f?

(i) $f(\emptyset) = 0$, i.e., the empty set has a zero measurement.

(ii) $f(A) \geqslant 0$ for every subset A. Note that it is possible to imagine $f(A) = 0$ for some sets A which are not empty. For example, if $S = R$, the real number system, and f measures the length of a set, then any set containing just one point has zero length.

(iii) $A \subset B \Rightarrow f(A) \leqslant f(B)$. We would certainly expect the larger set B to measure at least as much as its subset A.

(iv) If A, B are disjoint (i.e., they have no points in common) then $f(A \cup B) = f(A) + f(B)$. This agrees with our intuitive notion of the properties of weight, or length or volume.

(v) It is convenient to extend this notion to a sequence of sets $A_1, A_1, \ldots, A_n, \ldots$ for which no point is in two sets of the sequence. If we form the set E containing all those points of S which are in one of the A_i (E is called the union of the sets $A_1, A_2, \ldots$), then:

$$f(E) = f(A_1) + f(A_2) + \ldots + f(A_n) + \ldots$$

in the sense that the series converges to the correct limit.

HOW TO MEASURE THE SIZE OF A SET

This last property (v) is not as intuitively obvious as properties (i) to (iv). It means that if we divide a set E into a sequence of little bits which are all disjoint, then measuring each of the bits and adding should give the measurement of E. Because of condition (ii), all the terms of the series are non-negative, so we need not worry about the order in which they appear.

Any function f which has all these properties (i)–(v) is called a *measure*. The study of measures has proved extremely useful and interesting. Let us consider one or two interesting cases.

The simplest kind of space is one with a finite number of points, say:

$$S = \{x_1, x_2, x_3, \ldots, x_n\}$$

Think of each point having a measurement (see Fig. 9.7), so that:

$$f\{x_i\} = \mu_i \quad i = 1, 2, \ldots, n.$$

It is now simple to calculate f for any subset of S. Any set $A \subset S$ is the union of the single-point sets in it. Hence $f(A)$ is the sum of the numbers μ_i for those $x_i \in A$. For example:

$$f\{x_1, x_5, x_8\} = \mu_1 + \mu_5 + \mu_8$$
$$f\{x_2, x_4, x_6 \ldots\} = \mu_2 + \mu_4 + \mu_6 + \ldots$$

Here we can think of the points as having masses which are added together to give the mass of the set.

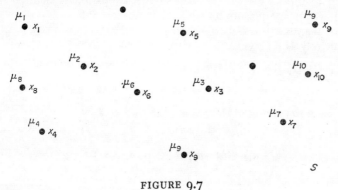

FIGURE 9.7

This example extends to a set S which is enumerable (see Chapter 8); for then we can write the points of S as a sequence:

$$x_1, x_2, x_3, \ldots, x_n, \ldots$$

in which each point appears once and only once. Now if $f\{x_i\} = \mu_i$ and:

$$\mu_1 + \mu_2 + \mu_3 + \ldots + \mu_n + \ldots$$

converges to a finite sum; then f is defined for any set $A \subset S$ by carrying out the possibly infinite addition of all the numbers μ_i for the points $x_i \in A$.

The situation is not so simple whenever S has a larger cardinal. The most important case of this situation is $S = R$, the set of real numbers. As usual we picture R as an infinite line. We would like to measure the 'length' of a set A of real numbers x with $a < x < b$. It is clear that the length of A should be $(b - a)$. This does not change if the set A is translated by adding the same number c to each point of A (see Fig. 9.8). Can we define a measure f on the real line which

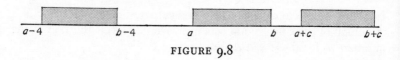

FIGURE 9.8

is translation invariant in this sense, and which assigns the normal length to an interval $a < x < b$?

Unfortunately, the answer is no—basically because there are too many possible subsets of R. However, it is possible to define f for a very large class of subsets with all these desirable properties. This particular f is called Lebesgue measure, and its definition can be given intuitively as follows.

Given any set A, cover it by a sequence of open intervals:

$$K_i = \{x \mid a_i < x < b_i\} \qquad i = 1, 2, \ldots$$

so that A is a subset of the union of these sets $\mathcal{J}_1, \mathcal{J}_2, \ldots, \mathcal{J}_n, \ldots$ The length of $\mathcal{J}_i$ is $(b_i - a_i)$ so the series:

$$(b_1 - a_1) + (b_2 - a_2) + \ldots + (b_n - a_n) + \ldots$$

gives an upper bound for the length of A. If we find that the series diverges for every covering of A, we say that A has infinite length. If it is possible to cover A by such a sequence in which the total length is finite, then the length of A is the greatest lower bound of the set of total lengths of covers.

HOW TO MEASURE THE SIZE OF A SET

Example 8. Find the length of the set Q of rational numbers. Note that there are rational numbers in every interval of R—in a sense they are spread all over the line.

We know that Q is enumerable, so that we can arrange it as a sequence:

$$r_1, r_2, r_3, \ldots, r_n, \ldots$$

for any given $\epsilon > 0$, let a_n, b_n be an interval of length $\epsilon/2^n$ centred at the nth rational r_n, so that:

$$a_n = r_n - \frac{\epsilon}{2^{n+1}}, \quad b_n = r_n + \frac{\epsilon}{2^{n+1}}$$

This means that $r_n \in K_n = \{x \mid a_n < x < b_n\}$, and the union of $\mathcal{J}_1, \mathcal{J}_2, \ldots, K_n, \ldots$ certainly contains every rational number. The total length is:

$$(b_1 - a_1) + (b_2 - a_2) + \ldots + (b_n - a_n) + \ldots$$

$$= \frac{\epsilon}{2} + \frac{\epsilon}{4} + \ldots + \frac{\epsilon}{2^n} + \ldots$$

$$= \epsilon \left[\frac{1}{2} + \frac{1}{4} + \ldots + \frac{1}{2^n} + \ldots \right]$$

$$= \epsilon$$

Thus Q can be covered by a set of intervals of total length ϵ. This can be done for every $\epsilon > 0$, so the length, or measure of Q must be zero. Although the rationals are everywhere spread along the line, the total length of the set of all rationals is zero—a somewhat surprising result. This is another sense in which most of the points on the line are irrational.

Example 9. In the language of probability theory we had a different kind of measurement of a set, which we called an event (Chapter 5). This measurement satisfied $P(S) = 1$ in addition to conditions (i)–(iv) for a measure. In fact the theory of probability becomes more tractable if we demand that the probability of an event is the sum of the probabilities of a sequence of events into which it can be subdivided, so that again it is usual to have condition (v). This means that the study of probability leads us to consider measures on S which assign the measure 1 to the whole space. Suppose now we take for S a segment of the real numbers from 0 to 1 and use the length function of Lebesgue measure to give the probability of subsets of S (see Fig. 9.9). This would be a good mathematical model for picking

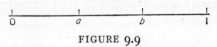

FIGURE 9.9

a point 'at random' on the segment S. The probability of picking an x satisfying $a < x < b$ is then $(b - a)$. What we have proved in Example 8 shows that if we pick a point at random between 0 and 1, there is probability zero that it will be rational.

BIBLIOGRAPHY

It would clearly be impossible to give an exhaustive reading list relating to a book such as this. Instead here is a brief list of two types of books which I enjoyed. Firstly, books of a general nature which cover a fair range of ideas; and secondly, books of a more specialist nature which you can consult if you wish to understand some particular topic more thoroughly. In most cases there are specific references to the books in the text: for each book I have added a brief description of its nature.

Books of a more general nature

R. Courant and H. Robbins, 'What is Mathematics?' Oxford University Press, 1941.
   An excellent account, at a level both wider and deeper than this book, of the true nature of mathematics.

L. Hogben, 'Mathematics for the Million', Pan Books, 1967.
   Develops mathematical thinking as a tool for the solution of problems in the real world. Interesting, but not for the novice.

E. Kasner and J. Newman, 'Mathematics and the Imagination', Bell, 1949.
   This book skates lightly over a very wide range of problems and ideas, usually presenting them in a humourous context.

F. Klein, 'Elementary Mathematics from an Advanced Standpoint', (2 Volumes; Vol. 1, 1924; Vol. 2, 1939), Dover Publications, distributed in the UK by Constable & Co.
   A classic, which shows how the development of mathematical research in the early part of the twentieth century is related to simple problems. Not for the novice.

E. P. Northrop, 'Riddles in Mathematics', Penguin, 1967.
   Explores a variety of apparent fallacies and contradictions.

D. Pedoe, 'The Gentle Art of Mathematics', Penguin, 1963.
   An interesting survey at a level similar to that of this book.

G. Polya, 'How to Solve It', Anchor Doubleday, distributed in the UK by Transatlantic Book Service.

A delightful little book which shows how a mathematician's mind works when presented with a problem.

H. Rademacher and O. Toeplitz, 'The Enjoyment of Mathematics', Princeton University Press, 1967.

This book explores the nature of mathematical thought by presenting specific problems and showing where they lead to.

Books providing further information about particular topics

J. C. Burkill, 'Mathematical Analysis', Cambridge University Press, 1962.

This presents in full detail the theory of convergence for series and sequences, and develops concepts related to functions of a real variable.

H. Davenport, 'The Higher Arithmetic: An Introduction to the Theory of Numbers', Hutchinson, 1968.

Develops a theory of the integers and examines the structure from an elementary standpoint.

W. Feller, 'An Introduction to Probability Theory and its Applications', Vol. I, Wiley, 1968.

An interesting book which does not shun hard problems.

E. Kamke, 'The Theory of Sets', Dover Publications, distributed in the UK by Constable & Co., 1950.

Develops the theory of both cardinal and ordinal numbers— without any previous knowledge assumed.

E. Landau, 'Foundations of Analysis', Chelsea, 1951.

This book really starts at the beginning, develops the integers from a set of axioms, and then constructs the real number system.

J. B. Roberts, 'The Real Number System in an Algebraic Setting', Freeman, 1962.

This book gives a systematic development of real numbers starting from the algebraic structure of the positive integers.

INDEX OF TERMS

Associative law applies if the manner of pairing terms makes no difference when carrying out the same operation twice   39

Base for a number system is a whole number k such that any other whole number can be expressed in terms of the digits 0, 1, 2, ..., $k-1$ 41

Bernstein's theorem relates the cardinals of sets which can be compared   148

Bijection is a function $f: A \to B$ such that to each point $y \in B$ there is a unique point $x \in A$ such that $y = f(x)$ 31

Binomial coefficient is the coefficient of $a^r b^{n-r}$ in the expansion of $(a+b)^n$ 48

Binomial theorem gives the result of raising a composite sum $(a+b)$ to the nth power   48

Buffon needle is the random experiment consisting of dropping a needle on a grid of equally spaced parallel lines   99

Cardinal number of a set is a measure of its size by the process of counting the elements   140

Cartesian product—another term for product set   22

Commutative law is an operation in which the order of the terms makes no difference   39

Compactness is a property of segments of the real line   75

Complement for a set A which is a subset of S, the complement A' or $S - A$ consists of precisely those elements of S which are not in A 7

Complementary set to A is another name for the complement of A 13

Completeness is the property of the set R of real numbers which makes precise the idea that there are no numbers missing   71

Composition of two functions $f: A \to B$, $g: B \to C$ is the function $h: A \to C$ given by first operating on an element x in A to give $f(x)$ and then operating on the result $f(x)$ to give an element in C 34

Contingency table analyses the occurrence of an event according to whether or not some other event has happened   85

Continuous function $f: I \to R$ is one for which the value $f(x)$ is close to $f(x_0)$ whenever x is close to x_0 75

Continuum is a name for the set of real numbers emphasizing that it has no gaps   144

Continuum hypothesis states that all infinite subsets of the real line have cardinal c or $\aleph_0$ 149

Convergent sequence $a_1, a_2, \ldots,$ $a_n, \ldots,$ is one for which all but a finite number of terms are within any prescribed distance of a fixed a 156

Convergent series is one for which the sequence of sums of the first n terms converges   161

Converse of a theorem is the result of interchanging the hypothesis and conclusion in the statement of the theorem. The converse of a valid theorem may or may not be valid   17

Co-ordinate is a specified element from an ordered set   22

Co-ordinate diagram is a way of representing the product set of two sets   23

Decimal representation of a real number expresses it exactly as an infinite sequence of digits (after the decimal point)   64

Decimal system is the device of using powers of ten to simplify counting and calculation   41

Defining sentence is the rule or method of deciding which objects belong to a given set   4

Denumerable set is one which has the same cardinal $\aleph_0$ as the set N of positive integers. There is a sequence $x_1, x_2, \ldots, x_n, \ldots,$ in which each element of the set appears once and only once   141

Diameter of a set is the largest distance separating two points of the set   90

Domain is the set on which a function is defined   31

Element is a general name for the objects which together make up a set   3

Empty set is the set containing no elements (also called Null set)   4

Euclidean algorithm is a theorem about the size of the remainder when you divide one integer by another   49

Euler's formula relates the numbers of edges, vertices, and faces for a polyhedron   119

Event is a subset of a probability space. When a random experiment is carried out it is completely determined whether or not the event occurs   80

Exclusive events are events such that no two of them can happen as a result of the same random experiment   81

Expectation of a random number is a kind of average value in a large number of independent repetitions   98, 113

Factorial of a whole number n is the result of multiplying together all the numbers $1, 2, 3, \ldots, n$ 48

Factorization of an integer into primes   53

Fair game is one in which the expectation of the net result of playing is zero   101, 114

Field is a set in which there are two operations—addition and multiplication—which can be carried out and always satisfy certain laws   62

Fixed point of a transformation $f: E \to E$ is any element x such that $x = f(x)$ 132

Function is a rule which assigns to each element of a set A a unique element in a set B 26

Goldbach's conjecture states that all even numbers can be

INDEX OF TERMS

expressed as the sum of two primes 54

Graph of a relation is the representation of the set $E \subset A \times B$ which defines the relation on a co-ordinate diagram 25

Greatest common divisor of two integers a, b is an integer d which divides both a and b and such that any integer x which divides both a and b must divide d 50

Identity function on a set E is the mapping $f: E \to E$ such that $f(x) = x$ for every x in E 36

Independent events A, B are such that the knowledge that A has happened makes no difference to the probability that B will happen 84

Induction is the process of proving a result true for every positive integer by proving that if it is true for the nth integer it is also true for the $(n+1)$th 43

Infinite set is one which contains more than m distinct elements for each integer m 148

Integer is another word for a whole number 38

Intersection of two sets A, B is the set of those elements which are in both A and B 9

Interval is a name for the set of real numbers x which lie between two fixed real numbers a, b 65

Inverse function Given a bijection $f: A \to B$, and a point $y \in B$ the unique element $x \in A$ such that $y = f(x)$ determines a rule or function $g: B \to A$ which is called the inverse function of f 31

Irrational number is any real number which is not representable in the form a/b 72

Jordan curve is a simple curve which returns to its starting point 130

Klein bottle is an example of a closed surface with only one side 129

Laws of large numbers make precise the idea that the relative frequency of an event gets close to the probability when the number of repetitions is large 95

Limit of a sequence which converges is the number approached as $n \to \infty$ 156

Lowest terms refers to a representation r/s of a rational number for which there is no positive integer dividing both r and s 61

Mapping is another word for function 26

Measure is a function defined on subsets satisfying certain conditions 169

Möbius band is an example of a surface with only one side 128

Monotone sequence is one such that all the terms grow larger as n grows, or all the terms grow smaller 158

Natural number is one of the numbers used in elementary counting 38

Necessary condition—involved in logical arguments; in English the clause expressing the condition is preceeded by 'only if' 17

Network is a collection of arcs and vertices 135

Null set is the set containing no

elements (also called Empty set) 4
Number line is a representation of the set of real numbers as points on an infinite straight line 61, 72

Onto A function $f: A \to B$ is said to map A onto B if every element in B is the image of at least one element in A 29
Ordered set E is a set with a relation $>$ on $E \times E$ such that for any $a, b \in E$ either $a = b$ or $a > b$ or $b > a$, and $a > b, b > c \Rightarrow a > c$ 42
Ordered pair consists of two elements in a specified order—chosen from the same set or from distinct sets 22

Pathwise connected set E is such that there is a curve lying in E which joins any pair of points $p, q \in E$ 127
Polyhedron is a figure bounded by plane surfaces 119
Power set of A is the set whose elements are the subsets of A 149
Prime number is a whole number which is not exactly divisible by any whole number other than itself and 1 16
Product set of two sets A, B is the set of ordered pairs (a, b) with $a \in A, b \in B$ 22
Proper subset of a set E is any subset of E other than the empty set and the whole set E 7

Random variable is a real number determined uniquely by the outcome of a random experiment 116
Random walk is the process of moving on the integers in which, at each stage, there are known probabilities of moving to the right or to the left 101, 105
Range is the set in which a function takes its values 31
Rational number can be represented as a fraction a/b where a, b are integers and $b \neq 0$ 60
Real number is any element of the ordered field R described in Chapter 4 56
Reductio ad absurdum is a method of proof by establishing a contradiction 16
Relation between sets A, B is given by a subset E of $A \times B$. For elements $a \in A, b \in B, a$ is said to be in relation ρ_E to B if and only if the pair $(a, b) \in E$ 20, 24
Relative frequency is the proportion of times an event happens in a sequence of repetitions 78
Ruin is the state reached in a gambling game when a player has no money left to stake 100

Sequence is a function defined on the set of positive integers 32
Set is a collection of objects (either in the real world or in our imagination) such that it is possible to decide whether or not any given object is in the set 3
Simple curve is a curve which does not intersect itself 126
Simple polyhedron is a polyhedron with no holes in it so that it can be deformed to a sphere 120
Space is a set containing all the elements which are relevant in a given context 5
Subset We say that a set A is a subset of a set or space S if every element in A is also an element of S 6
Sufficient condition—involved in logical arguments; in English the

clause expressing the condition is preceded by 'if' 17

Sum of a series $a_1+a_2+ \ldots +a_n + \ldots$ is the limit (when it exists) of the finite sums $S_n = a_1+a_2 + \ldots +a_n$ 165

Theorem is a statement about the logical consequences of certain assumptions 15

Topological property of a figure is any property which is preserved by topological transformations 125

Topological transformation of A to B is any bijection $f:A \to B$ which is continuous in both directions 125

Transformation is another word for function 26

Union of two sets A, B is the set of elements which are in either A, or B, or both 8

Upper bound of a set E on the line is a number b such that $x<b$ for all $x \in E$ 73

Venn diagram is a pictorial way of representing sets and how they are related 6

Well ordered set is a totally ordered set E with the property that every non-empty subset of E has a smallest element 43

Zeno's paradox is a specious argument which shows that motion is impossible 152

INDEX OF SYMBOLS

Symbol	Meaning	Page	Typical usage
$\in$	belongs to, or is an element of	3	$x \in A$ The element x belongs to the set A
$\{\ldots\}$	set consisting of elements between braces	3	$\{1, 3, 5, 7\}$ Set consisting of the numbers 1, 3, 5, 7
$\vert$	such that	4	$\{x \vert x = a^2 \text{ for an integer } a\}$ Set of x such that x is the square of an integer a
$\emptyset$	null set, or empty set	4	$\emptyset \subset A$ The empty set is contained in A
$\subset$	is contained in, or is a subset of	6	$A \subset S$ The set A is a subset of S
$\supset$	contains	6	$S \supset A$ The set S contains the set A
$A'(= S - A)$	complement of A	7	$(A')' = A$ The complement of A' is A
$\cup$	union	8	$A \cup B$ The union of A and B
$\cap$	intersection	9	$A \cap B$ The intersection of A and B
$\Rightarrow$	implies	15	$x > 8 \Rightarrow x > 0$ x larger than 8 implies that x is positive
ρ	is in relation to	20	$x \rho y$ the element x is in relation ρ to y
$\times$	cross	22	$A \times B$ is the set A cross B or the Cartesian product of A and B
(a, b)	the pair a, b	22	$(a, b) \in A \times B$ The pair (a, b) is an element of $A \times B$

INDEX OF SYMBOLS

Symbol	Meaning	Page	Typical usage
R	the set of real numbers	23	$x \Rightarrow \in R \ x^2 \geqslant 0$ x a real number implies that x^2 is greater than or equal to zero
R^2	Cartesian plane	24	$R^2 = R \times R$
$f(x)$	value of function f at x	27	$f(a) = b$ The value of function f at a is b
$f: A \to B$	f is a function defined on A taking values in B	28	$f: A \to B$ means that $x \in A \Rightarrow f(x) \in B$
(1, 1)	one-to-one	29	$f: A \to B$ is (1, 1) if $x_1, x_2 \in A$ with $x_1 \neq {}_2x \Rightarrow f(x_1) \neq f(x_2)$
$g \circ f$	g circle f	34	Given $f: U \to V, g: V \to W$, the composite function $g \circ f$ maps U to W
Id_E	identity function on E	36	$Id_E : E \to D$ is the mapping which sends each $x \in E$ to itself
N	the set N	38	N is the set of all positive whole numbers 1, 2, 3 . . .
|	divides	40	$a \| b$ if $ax = b$ for some x. We say a divides b
>	greater than, comes before	42	$a > b$ means that the element a comes before b
!	factorial	48	$4! = 4.3.2.1. = 24$
$\binom{n}{r}$	n above r	48	The binomial coefficient in the expansion of $(a+b)^n$
N_+	the set N_+	59	The set N of positive whole numbers together with the number zero
Q	the set of rationals	59	Every element in Z has a representation a/b, $b \neq 0$
Z	the set of integers	61	Z contains the negative integers as well as N_+
$[a, b]$	the interval a, b	67	For fixed real numbers a, b $[a, b]$ is the set of $x \in R$ with $a \leqslant x \leqslant b$

INDEX OF SYMBOLS

Symbol	Meaning	Page	Typical usage
P	probability	80	$P(H) = \frac{1}{2}$ If a fair coin is tossed, the probability of 'heads' is $\frac{1}{2}$
$\approx$	approximately equals	85	$\sqrt{2} \approx 1\cdot 41$
$E(\sqrt{\ })$	expectation of χ	114	A game is fair if $E(\chi) = 0$
$\sim$	has same cardinal as	140	$A \sim B$ means that there is a bijection $f : A \to B$
$\aleph$	aleph	141	$\aleph$ is the cardinal of the set N
c	continuum	144	**c** is the cardinal of R
(a, b)	interval a, b	144	Set of real numbers x such that $a < x < b$ is called an open interval
m	cardinal **m**	146	Any infinite set has cardinal $\mathbf{m} > \aleph_0$
$\|\|E\|\|$	cardinal of E	146	$\|\|N\|\| = \aleph_0$
2^A	power set of A	149	$\|\|2^N\|\| = \mathbf{c}$
$\to$	tends to	156	$\sqrt{n} \to \infty$ as $n \to \infty$
Lim	limit	156	$\lim_{n \to \infty} 2^{-n} = 0$
e	Euler's number	166	$e = \lim_{n \to \infty} (1 + \frac{1}{n})^n$